EXTRATERRESTRIAL LANGUAGES

EXTRATERRESTRIAL LANGUAGES

Daniel Oberhaus

THE MIT PRESS
CAMBRIDGE, MASSACHUSETTS
LONDON, ENGLAND

First MIT Press paperback edition, 2024
© 2019 Massachusetts Institute of Technology

This book was set in ITC Stone Serif Std and Futura Std by Toppan Best-set Premedia Limited. Printed and bound in the United States of America.

Library of Congress Cataloging-in-Publication Data

Names: Oberhaus, Daniel, author.
Title: Extraterrestrial languages / Daniel Oberhaus.
Other titles: Communication with extraterrestrial intelligence
Description: Cambridge, MA : The MIT Press, 2019. | Includes
    bibliographical references and index.
Identifiers: LCCN 2019001439 | ISBN 9780262043069 (hardcover :
    alk. paper)—9780262548649 (paperback)
Subjects: LCSH: Life on other planets. | Interstellar communication.
    | Space and time in language. | Language and languages. |
    Extraterrestrial anthropology. | Search for Extraterrestrial Intelligence
    (Study group : U.S.)
Classification: LCC QB54 .O225 2019 | DDC 576.8/39014—dc23 LC
    record available at https://lccn.loc.gov/2019001439

10 9 8 7 6 5 4 3

publication supported by a grant from
The Community Foundation for Greater New Haven
as part of the **Urban Haven Project**

Die Grenzen meiner Sprache bedeuten die Grenzen meiner Welt.

—Ludwig Wittgenstein, *Tractatus Logico-Philosophicus*

For my parents, Chris and Tenley

CONTENTS

Acknowledgments   xi

1 **A BRIEF HISTORY OF TALKING TO ALIENS** 1

Premodern METI   3
Modern METI   12

2 **FROM CETI TO METI**   19

Who Killed CETI?   20
Speaking of Communication   23
Do Aliens Speak English?   25
Extraterrestrial Cognition   32

3 **ALIENS ON EARTH**   37

Order and the Dolphin   40
Entropy and the Dolphin   49

4 **COSMIC COMPUTERS AND INTERSTELLAR CATS** 55

Language Corpora and ETAI   58
Cosmic OS   63
DNA as Executable Code   67

5   IS THERE A LANGUAGE OF THE UNIVERSE?   71

Will Extraterrestrials Understand Our Math?   77
SET(I) Theory   84
Embodied Extraterrestrial Intelligence   88

6   TOWARD A *LINGUA COSMICA*   93

Cosmic Calls   99
Lincos 2.0   104

7   HOW TO TALK IN SPACE   111

Physical Media   111
Microwave METI   119
OMETI   128

8   ART AS A UNIVERSAL LANGUAGE   135

The Conventionality of Images   137
Music of the Spheres   144

9   THE MANY FUTURES OF METI   155

Shouting in a Jungle   156
Is METI Scientific?   161
Profligate Transmissions   163
Who Speaks for Earth?   167

Appendix A: The Arecibo Message   171
Appendix B: The Cosmic Call Transmissions   179
Appendix C: Lincos   193
Appendix D: The Lambda Calculus and Its Application to
    Astrolinguistics   203
References   225
Index   247

ACKNOWLEDGMENTS

There is only a single author named on the cover of this book, but I'm afraid that is quite misleading. The book you hold in your hands could not have been written without the expert insight of Sheri Wells-Jensen, Yvan Dutil, Carl DeVito, John Elliott, Charles Cockell, Brenda McCowan, Eric Korpela, Richard Braastad, Seth Shostak, and Marlin Schuetz. I am indebted to you all. I would also like to thank Marc Lowenthal, Anthony Zannino, and Judith Feldmann of the MIT Press for their invaluable guidance (and patience) while shepherding this text from inception to publication. Finally, I offer a special thanks to my family, friends, and colleagues for their inexhaustible support and encouragement.

1

A BRIEF HISTORY OF TALKING TO ALIENS

In 1961, nine of the smartest individuals in the United States received a rather unusual letter in the mail. It consisted of a long string of binary digits and a short message: "Here is a hypothetical message received from outer space. It contains 551 zeros and ones. What does it mean?" Neither the sender nor his recipients knew it at the time, but this letter would later serve as the prototype for the first message for extraterrestrial intelligence ever broadcast into space. Its initial test run on Earth, however, was a total failure.

The prototype interstellar message was created by the planetary astronomer Frank Drake in the wake of a conference he organized at the National Radio Astronomy Observatory in Green Bank, West Virginia, where he had concluded Project Ozma, the first microwave search for extraterrestrial intelligence (SETI), only a few months earlier. The three-day conference was dedicated to assessing the viability of scientific SETI and was attended by a small cohort of leading physicists, chemists, biologists, and engineers who had demonstrated an interest in the possibility of extraterrestrial life. It was a landmark event that shaped the trajectory of SETI for decades to come. Yet in the months following the historic meeting, Drake realized that if

SETI were successful and detected a signal from outer space, this would raise a serious issue that had been neglected at the conference: how to design a response.

So he set about designing an experimental interstellar message that consisted of a string of 551 binary digits that could be arranged so that their bit-values formed pictures. The number 551 is semiprime, a design feature Drake hoped wouldn't escape the notice of an extraterrestrial—or his human test subjects. The number served as a sort of instruction manual for how to arrange the string of bits so that it formed a 19×29 bit array, which would reveal several bit images. These images depicted things like numbers and a human figure, but many of the images required a great deal of imaginative interpretation. Drake wanted to know whether there was any chance an extraterrestrial would understand the significance of his message, so he sent it to each of the attendees at the Green Bank conference as a test. If anyone could decipher the code, one would presume it would be the nine scientists who had given the most thought to the challenges of interstellar communication.

Drake received a single reply to his message. It was from Barney Oliver, the director of Hewlett Packard Labs, who responded with his own semiprime binary string. When Drake translated Oliver's reply into a bit array, he found that it contained a "simple and inspiring" message: a picture of a martini glass with an olive in it. Although Oliver had understood the format of Drake's message, he had failed to interpret even the simple numbering scheme coded in the message. The inability of the Green Bank attendees to decipher Drake's experimental interstellar message was not for want of intellect, however. Drake later sent the message to a few Nobel laureates, all of whom either failed to decipher it at all or arrived at incorrect interpretations. One physicist,

for example, interpreted the binary string as a close approximation of the quantum numbers that describe the arrangement of
electrons in an iron atom. It was only when Drake submitted the
message to a magazine for amateur code breakers that an electrical engineer from Brooklyn wrote to him and demonstrated that
he had correctly deciphered most of the message.

Given the difficulty that some of the brightest Earthlings
encountered when trying to decipher the bitmap, it seems
unlikely that an extraterrestrial intelligence would fare any better. Today, Drake's failed experiment in interstellar messaging
remains an instructive lesson for contemporary METI (messaging extraterrestrial intelligence) efforts insofar as it calls attention to the multitude of latent conventions that haunt human
cognition and communication. If an interstellar message has
any hope of being understood by an extraterrestrial recipient,
these conventions must be identified and excised from the messages and replaced with elements that can be presumed to be
universally understood by any intelligent mind. The design of a
universal communication system has vexed some of the greatest
minds in history, but it wasn't until relatively recently that the
technological means became available to put these systems into
practice by broadcasting them across the cosmos.

PREMODERN METI

Prior to the Enlightenment, the problem of extraterrestrial communication was mostly framed in ecclesiastical terms, as philosophers struggled to recreate the perfect language of God (Eco
1995). Nevertheless, a handful of Renaissance works directly
addressed the difficulties of communicating with extraterrestrial
life. Perhaps the most notable example is *Man in the Moone*, a

novel written by the English bishop Frances Godwin. Posthumously published in 1638 under the pseudonym Domingo Gonsales, the book recounts the adventures of the author, who is carried to the moon by geese and encounters an extraterrestrial race that speaks in a musical language. Gonsales notes that the "difficulty of that language is not to be conceived, and the reasons thereof are especially two: First because it hath no affinitie with any other ever I heard. Secondly, because it consisteth not so much of words and letters, as of tunes and uncouth sounds, that no letters can expresse. For you have few words, but there are consisting of tunes onely, so as if they list they will utter their minds by tunes without words" (quoted in Davies 1967). Godwin's novel is remarkable for anticipating the difficulties involved with interstellar messaging more than three hundred years before the first message was sent into the cosmos as well as the role of music in facilitating extraterrestrial communication.

By the nineteenth century, a handful of mathematicians began to develop programs meant specifically for communicating with extraterrestrials that were thought to exist on the moon or Mars (Ball 1901; Raulin-Cerceau 2010). The first scientific program for messaging extraterrestrials is widely attributed to the mathematician Carl Friedrich Gauss, who reportedly suggested creating a massive visual proof of the Pythagorean theorem in the Siberian tundra. This visual proof was to consist of a right triangle bordered on each side by squares and would be created by planting rows of trees for the borders and filling the interior of the space with wheat. Despite the minimalism of the design, Gauss's proposal was information rich. It would demonstrate that our species had mastered large-scale agriculture as well as basic mathematics, geometry, and logic. Gauss's idea appears to have influenced the Austrian astronomer Joseph Johann von Littrow, who later advanced his own plan for establishing contact with the moon-dwellers that involved digging massive trenches in the Sahara Desert in various shapes. These trenches would be filled with water and surfaced with kerosene, which would be set alight to send flaming geometrical messages to our cosmic neighbors. Suffice it to say, neither Gauss's nor von Littrow's outlandish communication schemes were ever put to the test.

Like non-Euclidean geometry and many of the other ideas that Gauss attempted to claim credit for, the attribution of this interplanetary communication system to Gauss is contested. Many nineteenth-century sources that mention the idea differ significantly in its details and none of them cite where Gauss had actually written about this plan (Crowe 1999). Still, personal letters indicate that the mathematician *was* fascinated by the idea of communicating with aliens that were supposed to live on the moon. Indeed, Gauss's desire to communicate with extraterrestrials was a motivating factor in his design of the heliotrope, a mirror that could be used to communicate over long distances using reflected light. In his design, Gauss planned to use an array of sixteen mirrors "to get in touch with our neighbors on the moon," a feat he estimated "would be a discovery greater than that of America" (Crowe 1999). Although we now know the lack of a lunar atmosphere precludes the possibility of even microbial life, Gauss was correct in his estimation of the importance of extraterrestrial contact.

By the end of the nineteenth century, enthusiasm about the prospect of establishing contact with extraterrestrials had reached a fever pitch in Europe, and Paris had established itself as the intellectual capital for interplanetary communication. In 1874, the eccentric French poet and inventor Charles Cros petitioned the French government for funding to build a giant mirror that would use focused sunlight to burn messages into the Martian and Venusian deserts as a means of communicating with the extraterrestrials that he believed inhabited those planets. Alternatively, Cros later suggested that mirrors could be used in conjunction with electric lights to establish an interplanetary communication network. He envisioned a Morse-code-style system that would allow Earthlings to communicate basic ideas about numbers to establish a common language with the Martians (Moore 2000). Around the same time, Nicolas Camille Flammarion, a French astronomer with a notable interest in the possibility of extraterrestrial life, also advocated using large mirrors to communicate with life on other planets. Rather than adopt Cros's telegraphic system, however, Flammarion envisioned an interplanetary messaging system based on large-scale arrays of reconfigurable electric lamps. According to Flammarion, the ability to change the shapes of the lights was critical because it would demonstrate to the Martians that the signal was artificial (Cerceau 2015).

Unsurprisingly, these exotic plans to contact extraterrestrials were ripe fodder for the humorists of the era such as the French playwright Tristan Bernard, who wrote a story about the first astronomers to receive a message from Mars. After receiving the message, the astronomers blanket the Sahara Desert with a reply that reads "I beg your pardon?" only to have the Martians reply, "Nothing." Vexed at the Martians for their terse answer, the astronomers write another message in the Sahara which reads "What are you making signs for then?" to which the Martians reply, "We're not talking to you, we're talking to the Saturnians" (Reddy 2011). Despite the satirical jabs, however, establishing contact with extraterrestrials was considered by many to be a legitimate pursuit, albeit one that depended on the patronage of wealthy xenophiles. In a testament to the strength of this belief in extraterrestrial life, the wealthy French socialite Anne Goguet established a prize for the first person to establish interplanetary communication in 1891. The contest was included in her will at the urging of her son Pierre Guzman, who was deeply interested in Flammarion's work on extraterrestrials. According to Goguet's will, the Prix Guzman would award 100,000 francs (equivalent to about $500,000 USD today) for the first demonstration of interplanetary communication. Notably, establishing contact with Martians wasn't eligible for the prize, as Goguet considered the existence of extraterrestrials on the Red Planet to be "sufficiently well known."

For better or worse, none of the early French proposals for interplanetary communication ever came to fruition, but they are remarkable for explicitly addressing many of the conceptual problems associated with modern extraterrestrial communication, even if many of the early plans focused on practical issues of interplanetary communication systems rather than the content of the messages. The first attempt at a complete program for interplanetary communication was designed in 1896 by the English polymath Francis Galton, who put forth a theory of extraterrestrial message construction based on Morse code. Inspired by the Mars craze that had swept Europe in the late nineteenth century after the Italian astronomer Giovanni Schiaparelli announced he had discovered canals on the planet's surface, Galton started toying with the idea of communicating with the Martians. Unlike the challenges associated with communicating with the blind or deaf on Earth, Galton observed, interplanetary communication does not have the luxury of feedback, which meant that "signals have to be devised that are *intrinsically* intelligible." Galton described his plans for extraterrestrial communication in an entertaining story about the first humans to receive Martian signals, which would consist of "minute scintillations of light proceeding from a single, well-defined spot on the surface of Mars." The Martian signaling pattern envisioned by Galton consisted of light flashes of three different durations that could be combined to form letters and words, much like Morse code. In Galton's fable, the Martian message begins with a series of lines to indicate the start of the message, before moving on to the notion of identity and arithmetical operators. Next, the message describes the "five principle planets" by their average distance from the Sun, rotation period, and circumference. With this foundation words consisting of groupings of three or more signals could be used for a "picture writing" to extend the system to more complicated concepts. Although it relied heavily on the conventions of Morse code, which presumably wouldn't be known to an extraterrestrial intelligence, Galton's system for communicating with extraterrestrials is remarkable for its recognition that an optimal extraterrestrial message would be self-interpreting, as well as for its systematic explanation of astronomical facts through a "language" of basic arithmetic and pictures. Indeed, Galton's idea of transmitting pictures was given a second life two decades later in a proposal that used the dots and dashes of Morse code as the basis of an interplanetary picture-messaging system. In this proposal, groups of twenty signals (either a dot or a dash) would serve as a line of an image, creating the equivalent of a bitmap (Nieman and Nieman 1920).

The idea of using mirrors to establish contact with Martians continued well into the early twentieth century (Mercier 1899; Pickering 1909), but Guglielmo Marconi's pioneering work on radio communication promised a more effective way of communicating across the cosmos. In 1899, a strange concurrence of events laid the foundation for modern METI, even if it wasn't recognized at the time. That year, Marconi managed to transmit a single letter via radio across the English Channel, while Nikola Tesla, infamous for his own forays into experimental radio technologies, recorded in his notes that he had detected a radio signal from Mars at his laboratory in Colorado. "I have observed electrical actions, which have appeared inexplicable," Tesla wrote in a letter to the New York Red Cross. "Faint and uncertain though they were, they have given me a deep conviction and foreknowledge, that ere long all human beings on this globe, as one, will turn their eyes to the firmament above, with feelings of love and reverence, thrilled by the glad news: 'Brethren! We have a message from another world, unknown and remote. It reads: one ... two ... three ...'" (Tesla 1900).

Despite Tesla's conviction that he had received the first interplanetary message—and it is worth noting that this alleged Martian message began by counting—it is likely that he had actually heard radio emissions from storms on Jupiter, an important lesson about the difficulties involved in distinguishing artificial and natural radio emissions (Corum and Corum 2003). Although there is no indication that Tesla ever received a follow-up message from Mars, he remained preoccupied with the idea of interplanetary communication for the rest of his career. By 1909, he had rejected the various French proposals for using mirrors to facilitate extraterrestrial contact in favor of using one of his wireless radio transmitters, which he had used to send "a powerful current around the globe." According to Tesla, the same principles could be harnessed to send a message to our planetary neighbors. Indeed, he claimed to have "already produced disturbances on Mars incomparably more powerful than could be attained by any light reflectors" during his experiments over the course of 1899 and 1900. "Electrical science is now so far advanced that our ability of flashing a signal to a planet is experimentally demonstrated," Tesla wrote in an op-ed for the *New York Times*. "The question is, when will humanity witness that great triumph. This is easily answered. The moment we obtain absolute evidences that an intelligent effort is being made in some other world to this effect, interplanetary transmission of intelligence can be considered as an accomplished fact. A primitive understanding can be reached quickly without difficulty. A complete exchange of ideas is a greater problem, but susceptible to a solution" (Tesla 1909).

The possibility of using radio to wirelessly contact inhabitants on other planets also captured the imagination of Marconi, who allegedly conducted several experiments to that end. In 1919, the *New York Times* reported that Marconi pioneered interstellar communication in 1909 when he sent a radio signal into the universe in the hopes that it would facilitate contact with extraterrestrial intelligence in other solar systems ("Radio to the Stars, Marconi's Hope"). If Marconi did in fact undertake this experiment with the intention of contacting extraterrestrials, it is unlikely that his transmitter was broadcasting at a high enough frequency to transverse the ionosphere, let alone powerful enough to reach even our closest stellar neighbor. Still, Marconi had no doubts about the feasibility of interplanetary communication. "In 10 years, probably much less, the world will be able to send messages to Mars directly and unhesitatingly, without a hitch or a stop or a word lost in space," Marconi said. "That it is possible to transmit signals to Mars I know as surely as if I had a gun big enough or powder strong enough to shoot there; more surely, in fact, for a gun might miss the mark, while my wireless message will strike the entire solar system without aiming" ("Ethergrams to Mars," 1906). In the 1920s, the American press began to circulate stories that Marconi had possibly detected intelligent radio signals from Mars and was actively attempting to establish communication with the planet. Marconi himself remained vague about whether he was attempting to communicate with Mars and doesn't appear to have directly referred to any radio experiments with the Red Planet beyond hypotheticals (Brown 2005). When asked about the theoretical possibility of communicating with extraterrestrials, however, Marconi acknowledged that it was technically feasible, but that the language barrier would be an obstacle to communication. "You see, one might get through some such messages as 2 plus 2 equals 4, and go on repeating it until an answer came back signifying 'Yes,' which would be one word," Marconi told one interviewer. "Mathematics must be the same throughout the physical universe. By sticking to mathematics over a number of years one might come to speech. It is certainly possible" ("Inter-Stellar Wireless a Possibility to Marconi," 1919).

In an event that prefigured contemporary fears about broadcasting interstellar messages, an op-ed was featured on the front page of the *New York Times* listing the follies of Marconi's proposal the day after an article ran about his alleged Martian experiments. The anonymous author of the editorial argued that since "quite possibly there are even yet more things in heaven and earth than are dreamed of in our philosophy, and it would be better to find them out in our own slow, blundering way rather than have knowledge for which we are unprepared precipitated on us by superior intelligences." The op-ed goes on to say that establishing communication with an extraterrestrial intelligence "would soon become annoying" should that intelligence take it upon itself to correct the errors in Earthling science and mathematics. "We may be all wrong about two and two equaling four, but to find it out too suddenly would give a jolt to all the process of human thinking," the article concludes. "Our knowledge of the universe is a poor thing, but our own" ("Let the Stars Alone," 1919).

MODERN METI

In 1932, Karl Jansky serendipitously observed radiation coming from the Milky Way at Bell Labs, thereby inaugurating the science of radio astronomy. By this point, the evidence that humans were the only intelligent species in the solar system was mounting, but Jansky's breakthrough turned the entire galaxy into fertile hunting grounds for extraterrestrial life. In the decades since Jansky's discovery, radio astronomers have yet to detect so much as a simple beacon indicating that there is intelligent life elsewhere in our galaxy, but this didn't stop scientists from proactively addressing the troubling problem foreseen by Tesla: how to exchange ideas with extraterrestrials once contact was established.

In 1952, the experimental zoologist Lancelot Hogben stood before the British Interplanetary Society and outlined his plan for a cosmic language called the Astraglossa. Although he later admitted that the project was a "*jeu d'esprit* for an evening's entertainment," he noted that it nevertheless "had an undertone of serious intention" (Hogben 1961). Indeed, Hogben's presentation marked the beginning of a "symbolic turn" in the field of interstellar communication, which continues to dominate the field to this day. Whereas many premodern METI proposals were iconic or derived from conventional communication systems like Morse code, the new generation of symbolic interstellar messages sought to create sophisticated systems that would allow the communication of not only mathematics but complex scientific, psychological, and sociological information. Given that an extraterrestrial recipient cannot be presumed to know anything about the conventions of communication systems on Earth, the new METI systems adopted self-interpretation as a guiding principle. The messages would have to be able to explain themselves, as it were, which would require beginning with knowledge that can reasonably be assumed to be universal, encoding this knowledge in a symbol system, and using this symbol system to build increasingly complex statements. Hogben's Astraglossa, for example, sought to build a radio lexicon and syntax by using numbers as a basis for the discussion of astronomical phenomena known to both the sender and receiver, such as the relative position of the planets in the recipient's solar system. In 1960, the same year that Frank Drake inaugurated modern SETI with Project Ozma, the first microwave survey of a nearby stellar system for signs of intelligent life, the Dutch mathematician Hans Freudenthal published *Lincos: Design of a Language for Cosmic Intercourse*. Although Freudenthal's system bears a remarkable similarity to Hogben's Astraglossa, neither appears to have been aware of the other's work prior to publication and Lincos is generally considered to be the first fully developed symbolic communication system for interstellar communication. A portmanteau of *lingua cosmica*, Lincos was conceived as a didactic exercise and reflected Freudenthal's pragmatic approach to teaching mathematics (Gravemeijer and Terwel 2000). Beginning with basic numerals, the text builds up a language that attempts to describe concepts from the physical sciences, as well as more abstract notions such as love and death. Freudenthal's *lingua* is unique insofar as its syntax relied heavily on symbolic logic yet rejected complete formalization. (For a complete discussion of Lincos syntax, see appendix C.)

Fourteen years after Freudenthal published *Lincos*, Frank Drake and Carl Sagan sent the first interstellar message from the Arecibo telescope in Puerto Rico to a star cluster approximately 22,000 light years from Earth. Rather than adopt Freudenthal's Lincos, which Drake considered to be a "rather risky method for interstellar communication, because it assumed that the recipients have brains and logic very similar to ours," Drake and Sagan instead chose what they considered to be a "simpler unambiguous method" (Sagan et al. 1978). The Arecibo message was modeled on the prototype message Drake designed in 1961 and consisted of 1,679 binary digits arranged as a rectangular bitmap. The resulting image depicts the numbers one through ten and the atomic numbers for the five elements that make up DNA, as well as the formulas for the sugars and bases in DNA nucleotides, a crude drawing of a human, a graphic representation of the solar system, and a picture of the Arecibo telescope (see appendix A for a more in-depth discussion of the contents of the Arecibo message). Although it was the most powerful radio transmission ever sent into space for the purposes of interstellar communication, the Arecibo message was more about demonstrating the technological capacity of new instruments on the Arecibo telescope than an earnest attempt at establishing interstellar communication (Sagan et al. 1978). Nevertheless, the message was optimized for intelligibility and transmission, so even though it is unlikely that aliens will ever receive it, the transmission is still considered a scientific attempt at interstellar communication (Zaitsev 2002).

Three years after this historic broadcast, two probes left Earth on a journey to the edge of the solar system. The Voyager missions launched about two weeks apart in 1977, each carrying a golden record designed as a sort of time capsule to represent life on Earth. Created by a committee chaired by Sagan, the records contain a diverse collection of visual and audio content, including 115 images from Earth, field recordings of everything from the ocean surf to bird song, musical selections ranging from Bach to Chuck Berry, an hour's worth of brainwaves collected from the committee's creative director Ann Druyan, a diagram of the hydrogen molecule, and a recording of Sagan laughing. The records were inspired by the small plaques that had been affixed to the Pioneer probes launched in 1972 and 1973, which depicted a nude man and woman, as well as a map of our solar system to show the origin of the spacecraft. The golden records carry far more information than the Pioneer plaques, but none of the craft are likely to ever be intercepted in deep space. Even if they are, it is unlikely that their contents will be understood, for reasons we will see later. Thus, as Sagan later remarked, although the Voyagers are unlikely to ever make contact with an extraterrestrial intelligence, the records still "provided us with a unique opportunity to view our planet, our species, and our civilization as a whole, and to imagine the moment of contact with some other planet, species, and civilization" (Sagan et al. 1978).

By the early 1990s, publicly funded SETI programs were dead thanks to the efforts of several US congressmen who objected to using taxpayer dollars to search for "little green men with misshapen heads," but this didn't stop research on the design of interstellar messages (Blum 1990). In 1990, the mathematician Carl DeVito and linguist Richard Oehrle published their research on a novel communication system that was based on the "fundamental facts of science," which demonstrated that it was possible to build a communication system capable of conveying a great deal of our scientific knowledge using basic arithmetic as a foundation. Shortly before the new millennium, two Canadian physicists broadcast the first scientific interstellar message in a quarter of a century from the Evpatoria radio telescope in Ukraine. On May 24, 1999, the first of two "Cosmic Call" messages were broadcast to a star 70.5 light years away in the constellation Cygnus (see appendix B for a detailed discussion of the design of these transmissions). The design of the message was directly inspired by Freudenthal's Lincos, although it departed from Freudenthal's design in some key ways. The full message consisted of twenty-three 127×127 bit "pages" that used a unique symbol system to convey a wide variety of topics ranging from basic arithmetic to the physical composition of Earth's crust. The following month, the same message was sent to three other stars located between 51.8 and 68.3 light years from Earth (Dutil and Dumas 2016).

In 2001, the first musical interstellar message was sent from the Evpatoria radar in the form of a theremin "concert for aliens." Two years later in 2003, a second, slightly modified Cosmic Call message was sent from Evpatoria to five different stars within forty-six light years of Earth. This second Cosmic Call transmission also included the binary code for a rudimentary chatbot named Ella. Created by the computer programmer Kevin Copple in 2000, Ella represented the epitome of natural language processing at the time. She can crack jokes, play blackjack, and if Copple gets his way, she will also be the first interstellar ambassador for Earth, assuming that the extraterrestrials on the receiving end of the message can figure out how to run the software. While this of course remains an open question,

Ella represented a dramatic and creative departure from typical interstellar message design and prefigured message designs based on artificial intelligence.

In terms of symbolic systems for interstellar communication, the most significant advance in the field in the past half century was the publication of *Astrolinguistics* by the Dutch mathematician Alexander Ollongren in 2013. The text is essentially a second-generation version of Freudenthal's Lincos that applies recent developments in computer science to the problem of interstellar communication. By repurposing the lambda calculus and the calculus of constructions, Ollongren designed a metalanguage that would allow extraterrestrial recipients to decipher natural language texts via applied higher-order logic. Aside from Lincos, it is the only other fully fledged language (that is, one consisting of a robust syntax and semantics) that has been designed for the sole purpose of communicating with extraterrestrials (see appendix D for a brief introduction to the lambda calculus and its application to Ollongren's Lincos).

From solar mirrors designed to communicate with Martians in Morse code to phonographic records and artificial intelligences meant to roam the cosmos for millennia, each interstellar communication system reflects the scientific knowledge and cultural sensibilities of the era that created it. The chosen medium for interstellar communication reveals a great deal about the technological sophistication of the transmitting civilization, but the real meat of a transmission is the information embedded in the message itself. In the following chapters, we will consider how philosophy, linguistics, mathematics, science, and art have informed the design of many of the interstellar messages described above.

2
FROM CETI TO METI

In 1971, dozens of astronomers and astrophysicists convened in a conference room at the Byurakan Astrophysical Observatory in Armenia for the first joint American and Soviet conference on communicating with extraterrestrial intelligence. American and Soviet scientists had met independently to discuss the topic in the early 1960s, but Cold War tensions made a joint meeting impossible at the time, a disheartening reality for those researchers contemplating extraterrestrial life in the cosmos. Early work had made it clear that a staggering range of scientific expertise would be necessary to adequately address the extraterrestrial question, but much of this expertise was cordoned off by the Iron Curtain. In this sense, the Byurakan conference was as much about extraterrestrials as it was about easing tensions between two global superpowers. As the Soviet radio astronomer Iosif Shklovskii remarked ahead of the conference, the prospect of communicating with extraterrestrials seemed dim if communication among terrestrial nations on the topic was prohibited (Sagan 1973).

Presiding over the landmark conference was a young Carl Sagan, who would make history three years later by sending the first interstellar message into the cosmos from the Arecibo radio telescope in Puerto Rico. The title and topic of the meeting was Communication with Extraterrestrial Intelligence (CETI), a use of language whose significance was not lost on Sagan. "The word CETI, which has been devised for this meeting, is I think appropriate in three different respects," Sagan noted in his opening remarks at Byurakan. "First, it is an acronym for Communication with Extraterrestrial Intelligence. Second, it is the Latin genitive for *whale*, which is of some interest to this discussion; the cetaceans are undoubtedly another intelligent species inhabiting our planet, and it has been argued that if we cannot communicate with them we should not be able to communicate with extraterrestrial civilizations. And finally, one of the two stars which was first examined by Frank Drake in Project Ozma, the first experimental undertaking along those lines, was Tau Ceti."

Sagan touched on a few of the more pressing topics when it comes to communicating with ETI, including the use of animals as extraterrestrial analogues and the choice of stellar targets. It is worthwhile, however, to consider whether "*communication* with extraterrestrial intelligence" is an appropriate term for the activity of sending information into interstellar space in the hopes of establishing contact with an extraterrestrial civilization.

WHO KILLED CETI?

During the three days in 1971 that Sagan and his colleagues occupied the Byurakan Observatory, they considered several proposals for beacons that could initially attract extraterrestrial attention in order to begin communication. In a sign of the times, the astronomer James Elliot, who would later discover the rings of Uranus, suggested that nuclear weapons could be a viable signaling mechanism. In his analysis of Starfish Prime, a 1962 test of a 1.4 kiloton nuclear warhead detonated at an altitude of 250 miles—the most powerful detonation ever in space—Elliot figured that the X-rays from this explosion could be detected at up to 400 astronomical units, or about ten times the distance of Pluto from the sun. While this is not nearly far enough to be detected by another solar system, Elliot suggested that simultaneously detonating all of Earth's nuclear weapons on the far side of the moon might do the trick. Based on an estimation of the size of US and Soviet nuclear stockpiles and an assumption that a device could be developed that could focus the detonation's resulting X-rays, Elliot calculated that a blast of this magnitude could be detected at up to 190 light years from Earth. It'd be a hell of a way to say hello, but extraterrestrials would have to be observing Earth at the time of blast, which Elliot conceded made it less than practical proposal. Aside from the exotic contact proposal broached by Elliot, the conference attendees focused mostly on microwaves as a mundane but eminently more practical interstellar communication medium. There were plenty of pragmatic concerns to be addressed when it came to radio transmissions, such as the ideal transmission frequency and choice of stellar targets, but one of the more pressing questions concerned the nature of the message's content. Among the various proposals for interstellar messages raised at Byurakan, one especially stands out. Marvin Minsky, one of the progenitors of the field of artificial intelligence, suggested it would be best to send a cat. Not a description of a cat, but the cat itself. As we will later see, behind Minksy's humor is a serious proposal, but the point here is that at the time of the Byurakan conference, actively attempting to estab-lish communication with extraterrestrials was considered a legitimate pursuit, and proposals for how this might be accomplished were as plentiful as they were fanciful.

Although the pursuit of communicating with extraterrestrials is far from dead, in the decades since the conference CETI has been subsumed by the *search* for extraterrestrial intelligence, or SETI. That the international community of alien hunters has swapped "communication" for "search" is indicative of a fundamental shift in its priorities. After an initial flurry of scientific interest in the problem of how we might establish contact with an extraterrestrial civilization, it increasingly seemed that the minuscule odds of selecting the correct stellar target out of the millions of possible candidates made the task of sending messages a fool's errand at worst or cosmic crapshoot at best. At its root, the problem was one of scale and a lack of information about the cosmic environment. Prior to the discovery of the first exoplanet in 1992, each solar system whose star fell within a certain range of parameters could be considered equally likely to host extraterrestrial life. At the time of the Byurakan conference, sending a message to a single stellar target was extremely resource intensive in terms of the amount of power required to produce a detectable signal across vast distance. Moreover, messages were limited to a narrow bandwidth centered on a single frequency, which also limited the amount of information that could be sent during transmission. Searching for signals, however, required far less energy. Furthermore, large radio telescopes on Earth could scan thousands of frequencies in a relatively wide swathe of the sky. Today, receivers on Earth are far more sophisticated than what Drake was using during the first microwave search in 1960, but the costs of transmitting are effectively the same.

Following Sagan and Drake's historical Arecibo transmission in 1974, the practice of so-called active SETI languished despite the proliferation of passive SETI experiments. Between 1974 and 1999, over fifty SETI searches targeting thousands of stars were conducted. During that same period, only a single transmission into space occurred. In 1986, the artist Joe Davis used MIT's Millstone Radar to send the sounds of vaginal contractions into space. This transmission lasted only a few minutes and was cut short by a US Air Force colonel after he learned the content of Davis's broadcast. In 1999, the Cosmic Call transmission from the Evpatoria radar in Ukraine became the first scientific interstellar transmission in a quarter of a century. After facilitating the Cosmic Call transmissions, the Russian radio astronomer Alexander Zaitsev realized that the term "active SETI" was inadequate for this rebirth of alien communication efforts. Instead he coined the term "messaging extraterrestrial intelligences"— METI. Compared to active SETI, which Zeitsev argued was transmitting only to attract extraterrestrial intelligence attention and compel them to send a message to Earth, METI pursues "a more global purpose—to overcome the Great Silence in the Universe, bringing to our extraterrestrial neighbors the long-expected annunciation 'You are not alone!'" (Zaitsev 2006). It is instructive to note that Zaitsev felt compelled to create a new acronym for this effort, rather than rehabilitate CETI, which highlights an important distinction between communication and messaging when it comes to interstellar broadcasts.

SPEAKING OF COMMUNICATION

In 1632, Galileo Galilei published *Dialogue Concerning the Two Chief World Systems*, a text mainly concerned with comparing the Copernican and Ptolemaic conceptions of the solar system. Yet the *Dialogue* also takes into consideration the multitude of "stupendous inventions" created by humans over the centuries. Towering over them all, in Galileo's estimation, is language, which endows humankind with a "means to communicate his deepest thoughts to any other person, though distant by mighty intervals of place and time" merely "by the different arrangements of twenty characters upon a page!" Nearly four hundred years later, this singular human ability is no less wondrous, both in respect to its uniqueness in the animal kingdom and its unparalleled ability to convey our "deepest thoughts" to others.

The claim that humans are the only animal to use language is controversial. There have been many different schemas for delimiting human language and animal communication. The linguist Charles Hockett, for instance, included language's infinite scope, creative use, capacity for displacement, and ability to refer to displaced referents among the sixteen "design features" that he claimed distinguished it from animal communications (Hockett 1966). Nevertheless, some of these ostensibly unique properties of human language have allegedly been found in the animal kingdom as well. The "dance language" of honeybees, for example, incorporates a limited version of displacement, and dolphins appear to be able to refer to absent objects (von Frisch 1967). Yet, as the linguist Noam Chomsky has noted on several occasions, the difference between animal communication and human language is still more fundamental insofar as "language is not properly regarded as a system of communication." Rather, Chomsky argued that language is a means of expressing *thought*, and although language can be used to communicate, this "is not *the* function of language, and may be of no unique significance for understanding the functions and nature of language" (Chomsky 2002). In this sense, language is better understood as a subset of communication, which is readily apparent when one considers the staggering variety of communicative systems. Someone's posture as they sit in a chair, their style of clothes, a cat's hiss, or a bird's plumage can all be considered modes of communication insofar as each of these instances conveys information to an observer. Still, most of us would probably hesitate to say that "body language" is anything other than nominally related to natural languages like French or English.

The distinction between CETI and METI now comes into sharper relief. Consider again Elliot's proposal that we might detonate the world's nuclear arsenals on the far side of the moon as a means of attracting the attention of an extraterrestrial intelligence. If we were to try this, it might communicate to an extraterrestrial observer of the blast that there is a civilization in the Milky Way that has harnessed nuclear fission, is able to traverse its local cosmic environment, and so on. Yet we likely wouldn't consider this to be a *message* to the extraterrestrial, since a message connotes the use of language, and hence, the conveyance of a thought. Given that METI is concerned with the creation of messages for extraterrestrial intelligences, its primary task is designing a linguistic system that is optimized for making the thoughts of its designers intelligible to an extraterrestrial subject.

DO ALIENS SPEAK ENGLISH?

There's a wonderful short story by Robert Silverberg (1974) that recounts the reveries of an acclaimed Professor Schwartz who finds himself enthralled with the idea of interstellar travel.

In his fantasies, Schwartz encounters a reptilian extraterrestrial race with "weirdly human, sad Levantine eyes" which is renowned for indulging in psychedelic drugs and strange ritual dances. "Schwartz had spoken with them several times," Silverberg writes. "They understood English well enough—all galactic races did; Schwartz imagined it would become the interstellar *lingua franca* as it had on Earth—but the construction of their vocal organs was such that they had no way of speaking it, and they relied instead on small translating machines hung around their necks that converted their soft whispered hisses into amber words pulsing across a screen." Silverberg's winking reference to Anglophone self-importance is humorous for its absurdity, but it also raises an interesting question about interstellar communication. Intuition suggests that any natural language on Earth would be insufficient for communicating with an extraterrestrial intelligence since our languages are dependent on contingencies of Earth's social history. Yet does that necessarily mean that an extraterrestrial wouldn't be able to understand them?

The task of communicating with an extraterrestrial intelligence bears a striking resemblance to a thought experiment posed by the philosopher Willard Van Orman Quine (1960) to demonstrate the indeterminacy of reference. In this thought experiment, a linguist is the first to contact a remote tribe. Given that neither the tribespeople nor the linguist has ever encountered the other's language, how are they to begin meaningful communication? Intuition suggests that finding a point of mutual reference would be a fruitful beginning for the linguist's translation efforts. One day, the linguist observes a rabbit dart out from a bush and one of the tribespeople points at the rabbit and says "gavagai." The linguist thinks he has made a breakthrough. Obviously, "gavagai" means "rabbit." Yet matters are not so simple. After all, "gavagai" could translate to "undetached rabbit part" or "rabbit stage," and pointing will not suffice to determine which of these competing meanings is meant by "gavagai." Consider an attempt to distinguish between "rabbit" and "undetached rabbit part" by pointing at various parts of the parts of the rabbit in isolation while asking "gavagai?" This does little to resolve the referent, since each time a rabbit part is indicated it also involves pointing at the whole rabbit. Although it doesn't resolve the indeterminacy of reference, Quine's practical solution to the linguist's conundrum involves creating a working system of translations for all the "grammatical particles and constructions" that allow us to individuate items in English: plural endings, pronouns, numerals, the "is" of identity, and so on. Through this translation process, the linguist eventually reaches a point where he can point to two rabbits to ask "Is this rabbit the same as that rabbit?" which will, in practice, set him on his way to determining whether "gavagai" means "rabbit," "undetached rabbit part," or "rabbit stage."

Obviously, such an intensive translation process is not going to work for interstellar communication, where extreme time lags between sending a message and receiving a response preclude the iterative feedback process employed by Quine's linguist. In this respect, the problem of interstellar communication more closely resembles a problem encountered by linguists who attempt to decipher ancient scripts in unknown languages such as Linear B. These texts don't point to anything outside of themselves, so determining the referents of the symbols through these texts alone is impossible. The meaning of Egyptian hieroglyphs, for example, would likely have remained inscrutable were it not for the Rosetta Stone, which mapped the symbols to a known language. If natural language is used for interstellar communication, there needs to be some way to link the arbitrary symbols that constitute the words of our natural languages to their referents in the world. Yet, as Quine argued, this is a potentially unresolvable problem, for the inscrutability of reference he described also holds for speakers communicating in the same language. Nevertheless, humans are able to learn a language and communicate meaningfully with other speakers. This suggests that a natural language could in principle be taught to an extraterrestrial intelligence, provided there was a mechanism to link a natural language text to its referents in the world, perhaps through a metalanguage or images. Examining the process of human language acquisition can thus reveal the traits an extraterrestrial intelligence must possess in order to reasonably presume they could be taught a natural language. This analysis, in short, requires considering human children as prototypical alien intelligences.

One of the most remarkable facts about natural language acquisition is that children can wield discrete infinity, understood as the "use of finite means to express an unlimited array of thoughts" (Chomsky 2002). Until the 1960s, accounting for this phenomenon was one of the most perplexing problems in linguistics. For the first half of the twentieth century, linguistics was dominated by a paradigm where language—in the sense of *langue*, the abstract system of rules that determines how a linguistic community uses language—was a social object that came to be known through exposure (Saussure 2011). The basic idea was that children are born as a blank slate and piece together the rules of language by abstracting from examples in everyday life. Yet this theory of language acquisition can't account for how children come to be able to generate and understand an infinite number of linguistic expressions if they have only been exposed to a relatively small subset of utterances in their day-to-day life. This insight was at the core of the so-called *cognitive revolution* in linguistics, which emphasized the neurocognitive basis of language.

The opening salvo in the second cognitive revolution was the theory of generative grammar, which explained how individuals could come to understand and produce an infinite number of utterances (Chomsky 1957). This theory posited that knowing a language amounted to the use of a recursive generative procedure and that this procedure could be formally defined for a given language. Most of us learn our first language without ever being explicitly taught the generative rules that make speaking that language possible. The question, then, is where do these generative procedures come from? Given that a child is equally capable of learning any language on Earth and that every human on the planet uses language, this suggests that our capacity for language is a biological endowment. The theory that each of us is born with a "language organ" is known as *universal grammar*. As its name suggests, universal grammar is concerned with properties of language that can be found in every language. Importantly, universal grammar overturns the empiricist *tabula rasa* hypothesis, according to which the human brain is born a blank slate that comes to use language by collecting examples. Instead, the human brain must already be endowed with a highly structured system that makes the acquisition of any given language possible.

Theories about the relationship between universal grammar and particular grammars have evolved over the decades. By the 1970s it was clear that specific languages were far more similar than previously thought. Thus, universal grammar came to be defined as a system of parameters, and individual languages were defined by unique parametric values within this system. Beginning in the 1990s, Chomsky and his acolytes refined the principles and parameters theory with the *minimalist program*, which sought to answer the question of why the faculty of language has the properties it does. The strong minimalist thesis proposed by Chomsky (2000a) suggested that the human language faculty is an optimal solution to externally imposed conditions that must be satisfied for language to be used at all. It is important to note that "external" as used here implies systems that are external to the faculty of language (e.g., the sensorimotor system), not the person. This view takes the mind to be a collection of modules with specific properties, which are implemented on a plastic neural substrate. For the faculty of language to be useable at all, it must be able to interface with other systems of language (e.g., the sensorimotor and conceptual/thought modules). Since these modules each have their own unique properties, they impose "minimum design specifications" on the faculty of language that are necessary to allow these modules to interface with one another and for the individual to be able to vocalize thoughts (Chomsky 2000). The minimalist program seeks to elucidate the nature of these minimum specifications and whether these specifications are also an optimal solution to the constraints placed on the faculty of language.

Importantly, these cognitive modules are all implemented on a biological substrate consisting of billions of neurons, which suggests that there should be physical signatures corresponding to the unique features of the faculty of language. When Chomsky first theorized about generative grammar in the late 1950s, the technology to produce high-quality images of the brain was still three decades in the future. Now, however, neurolinguists have access to sophisticated brain-imaging technologies that provide a window to the functioning brain in real time, enabling them to probe the relationship between neurobiology and language. There is strong evidence to suggest that one of the defining features of all natural languages is a hierarchical, recursive syntax (Hauser, Chomsky, and Fitch 2002). At least three different experiments have used neuroimaging techniques to identify the physical signature of this unique language feature by exposing humans to sentences with both normal and garbled syntax (Embick et al. 2000; Monti, Parsons, and Osherson 2009; Moro et al. 2001). The results of these experiments were unambiguous: the recursive and hierarchical syntax in natural language is distinctly represented in our brain activity. In addition to bolstering the idea that language structure is biologically determined, Moro (2016) notes that these experiments also "dismantle a popular conviction that languages are forms of software running on a passive hardware base; if anything, they are the expression of the hardware activity, as if flesh became language, *logos*."

The second cognitive revolution in linguistics, and the mounting body of evidence from neuroscience that backs it up, has serious implications for METI since "the same structures that make it possible to learn a human language make it impossible for us to learn a language that violates the principles of universal grammar. If a Martian landed from outer space and spoke a language that violated universal grammar, we should simply not be able to learn that language the way that we learn a human language." Instead, we would have to "approach the alien's language slowly and laboriously—the way that scientists study physics, where it takes generation after generation of labor to gain new understanding and to make significant progress" (Chomsky and Gliedman 1983). Insofar as the faculty of language is linked to the structure of our brain, interstellar messages that contain natural language presume that the extraterrestrial recipient possesses a functionally equivalent neurobiology. Moreover, if the characteristics of our faculty of language are minimally determined by the demands of external systems, then the elucidation of these minimal features will help clarify other latent assumptions about the nature of the extraterrestrial intelligence. For example, if an extraterrestrial has a radically different sensorimotor system, then this could alter the nature of its faculty of language by altering its minimal design requirements. An interesting question, in this respect, is what aspects of the faculty of language are necessary for language as such, and which are contingencies imposed by other sensorimotor systems.

EXTRATERRESTRIAL COGNITION

The notion that aliens may think like us is justified on the grounds that we are both subject to similar environmental constraints. Not only are the laws of physics the same throughout the universe, but all life is subject to resource scarcity. Even if an extraterrestrial intelligence managed to colonize its solar system or build Dyson spheres to efficiently harvest energy from its host star, thereby greatly increasingly its energy and material resources, these resources are still finite. It has been argued that technological progress allows resources to be considered effectively unlimited, but this presumes that technological progress always outpaces resource depletion owing to economic pressures (Simon 1981). Since technical solutions cannot be guaranteed, however, one can reasonably suppose that all extraterrestrial civilizations will carefully manage their available resources. This argument from economy also lends strong support to the notion that ET's intelligence will be familiar to us. This was first recognized by two of the pioneers of machine learning, John McCarthy and Marvin Minsky. They realized that the search for extraterrestrial intelligence shared many similarities to their own search for artificial intelligence, insofar as it was the "attempt to study intellectual mechanisms as independently as possible of the particular ways intellectual activity is carried out by humans" (McCarthy 1974). This research paradigm was predicated on the idea that "the mechanisms of intelligence are objective and are not dependent on whether a human being or a machine or an extraterrestrial being is doing the thinking." Instead, the methods of intelligence are determined by the nature of the problem at hand. For example, the ability to play games like chess and Go requires mastering processes like searching for solutions and breaking a situation down into constituent elements. Although humans created these games, the fundamental processes involved in playing the game were successfully implemented in computers, which not only learned how to play the games but surpassed the abilities of their human opponents (Silver et al. 2017).

Extrapolating from the objective mechanisms of intelligence and a universal principle of economy, Minsky (1985) argued that we will be able to converse with an extraterrestrial intelligence because they will think like us. If all intelligent creatures are faced with the same fundamental problems (i.e., restraints on space, time, and materials) and the methods of intelligence are determined by the nature of the problem at hand, Minsky reasoned that extraterrestrial intelligences will arrive at solutions similar to our own, namely symbolic systems for representing these problems and processes for manipulating those systems that can also be described symbolically. The reason extraterrestrial intelligences will develop symbolic systems is because it is economical to do so. Strong solutions to the problems that emerge from restraints on space, time, and materials require the ability to effectively manage mental and physical resources. Minsky argued that this ability can emerge only if these resources are represented symbolically, which allows for the efficient transmission and accumulation of knowledge. In principle, the objective nature of intellectual processes means a student is capable of rediscovering all of physics from scratch, but that would be quite inefficient. Likewise, it would be wildly impractical if we were obliged to conjure a physical piece of fruit or a rocket engine whenever we wished to discuss an apple or interplanetary propulsion. Of course, it is possible to argue that even if extraterrestrial intelligences have symbolic systems, they may be so different from our own that we could never understand one another. Minsky rejected this argument by appealing to his sparseness principle, which posits that there are special ideas that "every evolving intelligence will eventually encounter" because they are simpler than other ideas that result in the same product. Two examples of these special ideas are numbers and arithmetic, which is why they may serve as a strong basis for an interstellar message.

Minsky made an even stronger claim, however, when he argued that many aspects of human language may be "inescapable" and likely to be found in extraterrestrial symbol systems as well. He argued that certain fundamental parts of our language, particularly the ideas of "thing" and "cause," reflect our mental processes, which were developed in our prelinguistic ancestors. Even if those ancestors didn't have language, they did have mental representations of objects and differences between those objects. The notion of a "thing" or "object" implies an enduring collection of properties and relations between those properties. This effectively delineates the thing from everything else, allowing the object to be recognizable even if, for example, it is moved to a different location. Importantly, the notion of "thing" applies not just to physical objects, but also to concepts and other immaterial entities (e.g., a song). The idea of "thing" is critical to the idea of "cause," for if there were no things it would appear as though everything that happens depends on everything else that happens. The world is an unfathomably complex place, but the concepts of "thing" and "cause" allow us to create substructures that can deal with the localized effects of things without having to consider the whole of everything else. An extraterrestrial without these fundamental notions of "thing" and "cause," which absorbed the world as one, undifferentiated whole, wouldn't be able to develop the necessary physics to build a radio telescope because everything would seem to be causing everything else. Moreover, when it comes to communication, the idea of a "thing" also endows humans with the ability to embed one clause within another clause. Minsky argued that this is the real key to our intelligence insofar as it allows us to represent prior thoughts as objects of another thought, a mental trick that manifests as recursion in natural language. "Unless aliens do that too," Minsky wrote, "they cannot turn their thoughts upon the prior products of their thoughts. Without this trick of turning symbols on themselves, one cannot have general intelligence—however excellent may be one's repertoire of other skills."

Thus, it would appear that the economy born of certain universal restraints—the laws of physics and the scarcity of resources—gives us good reason to suppose that extraterrestrial cognition will be sufficiently similar to our own to allow meaningful communication to take place. Of course, this is just a hypothesis, but so far no system for interstellar communication has been developed that doesn't assume at least some similarity between human and extraterrestrial mental processes. Even if it should be possible to communicate with extraterrestrials in principle, however, this points to the question of what form this communication should take. To clarify this issue, we will look at nonhuman communication systems on Earth as prototypical interstellar communication systems.

3

ALIENS ON EARTH

Modern SETI properly began in April 1960 with Project Ozma, when the twenty-nine-year-old Cornell astronomer Frank Drake trained the eighty-five-foot radio telescope at the National Radio Astronomy Observatory in Green Bank, West Virginia, toward Tau Ceti and Epsilon Eridani, two stars eleven light years away and approximately the same age as our Sun. For two hundred hours spread over the course of three months, Drake used the telescope to scan for signs of intelligent life around these two stars. The radio telescope was tuned to 1420 MHz, the radiation frequency of neutral hydrogen, which was considered the most likely frequency for an extraterrestrial broadcast because it is the most abundant element in the universe. Indeed, it seemed as though nature had come readymade with its own interstellar telecommunications system. Drake set up the radio telescope with two feeds so that the telescope's beam could alternate between targeting the stars directly and the empty sky beside the stars. Alternating between these two feeds was a way of filtering out terrestrial interference, which would appear in both feeds, whereas an extraterrestrial signal would only appear in one.

What might this extraterrestrial signal look like? Drake hypothesized that a signal would likely consist of a basic counting pattern, perhaps the first few primes. The physicist Edward Purcell, a Nobel laureate and the first to discover the radio emissions of neutral hydrogen, thought the signal would be even simpler and consist of a basic on-off pattern. Although Tau Ceti and Epsilon Eridani aren't the closest stars to Earth, at the time they were considered the closest to Earth that had any chance of supporting life. To have found intelligent life in the closest habitable solar system would either have been a remarkable coincidence or meant that intelligent life in the universe was far more common than anyone had assumed. Yet by the time Drake shut down Project Ozma in July, no signal—primes or otherwise—had been detected. This raised a serious question for the future of SETI: just how common is intelligent life in the universe?

Shortly after the conclusion of Project Ozma, the National Academy of Sciences requested that Drake meet with other scientists to determine the future of SETI—or, for that matter, whether SETI should have a future at all. The three-day meeting was scheduled to begin November 1, 1961, at Green Bank, and a few days ahead of the conference Drake pieced together an agenda that would address "all elements needed to predict the difficulty of detecting extraterrestrial life." The list he came up with included the average rate of star formation, the percentage of stars hosting planets, the average number of habitable planets per star, the percentage of planets where life emerges, the percentage of planets with life where intelligence evolves, the percentage of intelligent societies capable of interstellar communication, and the number of years a civilization is sending out detectable signals (Drake and Sobel 1992). Drake realized that multiplying the value of these items together resulted in a rough

estimate of the number of communicating extraterrestrial civilizations in the Milky Way—and with this realization, the famous
Drake equation was born. As for the conference attendees, Drake
worked with the National Academy of Sciences space science
board officer J. Peter Pearman to compile an invite list of "every
scientist [they] knew who was even thinking about searching
for extraterrestrial life." This turned out to be only ten people—
including Pearman and Drake. Although the list was short, the
invitees were a "who's who" of scientists doing cutting-edge
work in their respective fields. There was the chemist Melvin
Calvin, who was notified during the three-day conference that
he had won a Nobel Prize; the young planetary scientist Carl
Sagan; ham radio enthusiast and electronics entrepreneur Dana
Atchley; the astrophysicist Su-Shu Huang, who had coined the
term "habitable zone"; the inventor Barney Oliver, the director of Hewlett Packard labs who would later author NASA's first
report on SETI; Otto Struve, the director of the Green Bank
National Radio Astronomy Observatory, who was one of the first
scientists to attempt to estimate the prevalence of extraterrestrial
life in the universe; and the physicists Giuseppe Cocconi and
Philip Morrison, who authored the first scientific paper on the
feasibility of SETI in 1959. As Drake and Pearman compiled their
guest list, Drake joked that "all we need now is someone who has
spoken to an extraterrestrial," but Pearman, ignoring the humor,
had a suggestion: John Lilly.

At the time of the Green Bank conference, Lilly was working
at the Communication Research Institute in the Virgin Islands,
where he was making an earnest attempt to communicate with
dolphins. The year after the Green Bank conference, Lilly gained
widespread recognition for his work through the publication
of *Man and Dolphin*, in which he argued that dolphins may be

as intelligent as humans and that communicating with them should be possible. Lilly ended up going to great lengths to speak to dolphins, including the questionable practice of injecting his cetacean subjects with LSD, but his attempts at interspecies communication were never successful. In 1961, Lilly's scientific legacy hadn't yet been tainted with the gross ethical missteps that would come to define his career and as Sagan (1973) noted, "there was a feeling that this effort to communicate with dolphins was in some sense comparable to the task that will face us in communicating with an intelligent species on another planet, should interstellar radio communication be established." Indeed, Lilly's presentation at the Green Bank conference on dolphin communications made such an impression on the other attendees that the group adopted the informal moniker "The Order of the Dolphin." Although Sagan later reneged on his belief that Lilly's work held the key to understanding or designing interstellar communication, Lilly's research on interspecies communication profoundly shaped research directions of METI by positing interspecies communication as a model of interstellar communication.

ORDER AND THE DOLPHIN

In *The Mind of the Dolphin*, published a few years after the Green Bank conference, Lilly remarked that "currently we are faced with other species possibly as intelligent as we are," but "we do not yet recognize their intelligence." Lilly argued passionately that if we are to have any hope of understanding an extraterrestrial intelligence, then we must first solve the difficulties involved with communicating with nonhuman intelligences on Earth.

The Order of the Dolphin's hunch that the problems of inter-species communication are analogous to the problems faced in interstellar communication seems intuitively correct: here we have two radically different organisms each with their own species-specific mode of communication that are attempting to convey information to one another. Yet this analogy is inadequate for two reasons. First, it assumes that language is merely a form of communication. Second, it presumes that at least some animal communication systems—particularly those of dolphins and chimpanzees—are not essentially different from human languages. As Chomsky and others have argued elsewhere, human language can be used for communication, but this is apparently not its primary purpose. Natural languages are mostly used internally to order thoughts and only relatively infrequently to convey these thoughts to others. Therefore Chomsky refers to natural languages as *internal languages*, or *I-languages*, which must be distinguished from language as the *computational system* consisting of a small set of discrete rules—the universal grammar—discussed in the previous chapter. There is a nearly endless variety of communication systems, ranging from manner of dress and avian plumage to body "language" and the chemical secretions of plants. Although natural languages can fulfill these communicative functions, they cannot be reduced to them.

For Lilly, the activities of dolphins indicated that they were capable of thought, and their vocalizations suggested that these thoughts were being conveyed to one another through a type of rudimentary language, which Sagan once referred to as "dolphinese." Contemporary research suggests that Lilly overestimated the complexity of dolphin language, but his thoughts on the matter are still instructive for interstellar messaging insofar as they clarify the boundaries of language. Lilly defined communication as "the exchange of information between minds," which involved a certain degree of abstraction insofar as "information does not exist as information until it is within the higher levels of abstraction of each of the minds and computed as such. Up until that point at which it becomes perceived as information it is signals." Consider the case of pulsar CP1919 that was discovered at the Arecibo Observatory in 1967 and was mistakenly believed to be an extraterrestrial signal. In this case, we have a signal—electromagnetic radiation emanating from a certain region of the sky—but its information content is ambiguous. It could convey information about a rapidly spinning star, or it could convey the presence of an extraterrestrial civilization. It is only after more data were collected that scientists could abstract from this data to conclusively determine that it was a star, not an extraterrestrial beacon. Yet if this signal *had* turned out to be an extraterrestrial beacon that also contained a message, there would be still *another* level of abstraction beyond the information contained in the signal. The next level of abstraction beyond pure information involves deciphering the symbolic system or language coded in the signal in order to understand the *message* contained therein. By reducing language to mere communication, Lilly missed a crucial insight: communication systems only convey information, whereas language is also able to convey thought.

This was an insight recognized at least as early as 1859, when Charles Darwin remarked that "a long and complex train of thought can no more be carried on without the aid of words, whether spoken or silent, than a long calculation without the use of figures." This insight has profound implications for interstellar communication. If animals are indeed capable of language, then they are by extension capable of complex thought and can be considered intelligent in much the same way that we would describe humans as intelligent. Likewise, when we speak of "extraterrestrial intelligence," we are tacitly implying that the extraterrestrial has a capacity for language, and hence, for complex thought. Clearly, Lilly was operating on the assumption that dolphins were capable of complex thought and this intelligence manifested in their use of a species-specific language that was only waiting to be deciphered.

The notion of dolphins as an extraterrestrial analogue has haunted METI ever since the Green Bank conference (Fleury 1980; Herzing 2010), and not without reason. METI presupposes an extraterrestrial recipient that is intelligent and endowed with a faculty of language, and the vast repertoire of dolphin clicks and whistles do seem to bear many of the hallmarks of human language. For example, dolphins have demonstrated the ability to refer to one another with unique sounds (Janik 2000), the capacity for displaced reference (Herman and Forestell 1985), and the ability to understand novel instructions consisting of "sentences" that use up to five words (Herman 2010). When it comes to interspecies communication with nonhuman primates, things are quite similar. Bonobos are capable of displaced reference (Lyn et al. 2014), gorillas have demonstrated the understanding and creative use of over a thousand symbols in sign language (Patterson and Cohn 1990), and chimpanzees are perhaps the only nonhuman member of the animal kingdom to have demonstrated a capacity for basic symbolic addition (Dehaene 2011).

These experimental results ostensibly reinforce the idea that the "basic computational biological prerequisites for human language, including sentence and discourse processing, are already present in nonhuman primates" (Bornkessel-Schlesewsky et al. 2015). This view rejects the Chomskyan notion "that a more elaborate and qualitatively distinct computation mechanism (i.e., discrete infinity produced by recursion) is required for human language." Rather, "the ability to combine two elements A and B in an order sensitive manner to yield the sentence AB forms the computational basis for the processing capacity ... in human language." Yet others argue that this fundamentally misrepresents how human language (as a computational system) works. In human language, the linear order of elements in a sentence is not what makes it intelligible; rather, it is the hierarchical structure that is important. Consider an example given by Berwick and Chomsky (2016) to illustrate that linear order is insufficient for human language: the difference between *birds that fly instinctively swim* and *instinctively birds that fly swim*. The first sentence is ambiguous insofar as "instinctively" may modify either "fly" or "swim." In the second example, however, "instinctively" can only modify "swim." Yet "instinctively" occurs closer to the word "fly" than to "swim" in the sentence. Linear order cannot account for why "instinctively" modifies the more distant verb, but if one looks at the hierarchical structure of the sentence things are clearer. In terms of *structural* distance, "instinctively" is closer to "swim" than "fly." This reinforces Chomsky's thesis that human language syntax has at least three unique properties: "(1) human syntax is hierarchical, and is blind to considerations of linear order, with linear ordering constraints reserved for externalization; (2) the particular hierarchical structures associated with sentences affects their interpretation; and (3) there is no upper bound on the depth of relevant hierarchical structure." These properties have not been found in the communication systems of nonhuman primates or any other animals. This suggests that human language is in fact qualitatively different from other animal communication systems and, according to Berwick and Chomsky, makes "the nonhuman primate brain a poor candidate for modeling many aspects of the human language."

Still, the notion persists that nonhuman intelligence tends to converge toward human-like intelligence. Indeed, there is a preponderance of evidence to suggest that "convergence is the norm" (Coe, Palmer, and Pomianek 2013). The fact that the eye appears to have evolved independently several times on Earth, yet in each case evolved slightly differently because of the peculiarities of an ecological niche, is a well-worn example of convergence at work (Mayr 1985). Extrapolating from this, it is possible to argue that "the number of evolutionary end points is limited. ... What is possible has been arrived at multiple times, meaning that the emergence of the various biological properties is effectively inevitable" (Morris 2003). The controversy arises, however, as soon as human-like intelligence is considered. It is difficult to dispute that "high intelligence" and complex brains have evolved independently on Earth several times (Roth 2015), but it is equally difficult to account for why humans alone developed symbolic communication systems. Although some research suggests that animals develop large, complex brains to foster increasingly complex cognition (Jerison 1985; Marino et al. 2007), it's possible that complex animal brains are not actually indicative of any special intellectual capacity (Manger 2013). The inability to explain the uniqueness of language may also stem from an outdated theory of evolution where advantageous traits always emerge through a series of small variations in individuals over millions of years (Berwick and Chomsky 2016).

Modern evolutionary theories have abandoned this deterministic approach to evolution in favor of an evolutionary paradigm defined by stochastic (randomly determined) processes. On the classic deterministic view, nonhuman primates *must* have some aspects of the language faculty because they are our closest relatives. Yet modern theories of evolution allow for developmental "leaps" in evolution, which means that it is consistent that humans possess a language faculty while nonhuman primates do not even possess a rudimentary version of this mechanism.

When Sagan went to visit Lilly's dolphin research laboratory on the isle of St. Thomas, he was impressed by Lilly's attempts to establish communication with his cetacean subjects but couldn't help notice the one-sided nature of the conversations. "It is of interest to note," Sagan later reflected on his time at Lilly's lab, "that while some dolphins are reported to have learned English—up to 50 words used in correct context—no human being has been reported to have learned dolphinese." Sagan's incisive observation is as true today as it was when Lilly was injecting his dolphins with LSD in the 1960s and, despite its wit, offers a profound observation about the limits of interspecies communication as a model for interstellar messaging. Successes in communicating with nonhuman primates and cetaceans have all involved teaching symbolic systems to the animals, but even after all of this, we still cannot understand what the animals might be communicating among themselves. This is not for lack of effort, however. Since 1985, Denise Herzing and her colleagues have traveled to Bahamas once a year to study a group of around two hundred spotted and bottlenose dolphins (Herzing 2010). For about three months out of the year, the same group of dolphins congregates in the shallow waters north of the Grand Bahama Island, which allowed Herzing and her colleagues to document the behavior of individuals over the course of three generations, as well as record audio and visual data about the dolphins. Over time, the dolphins became accustomed to the presence of the researchers and would interact with them while they were in the water. In 1997, Herzing deployed a prototype interface that she described as "essentially an underwater keyboard that labeled objects, actions and locations, for dolphin access with a visual as well as acoustic signal. The acoustic signal design included frequency modulated whistles that were outside the normal dolphin repertoire of whistles but within their abilities to mimic." An initial test of the device revealed that the dolphins demonstrated active interest in the device after being exposed to the tones, and both species of dolphins interacted with the system during sessions. In 2012, Herzing deployed a new version of this interface called Cetacean Human and Telemetry (CHAT), which is effectively a two-way underwater computer that can receive dolphin signals and generate responses with a small speaker (Herzing 2014). The sounds that could be generated by the device are outside of the dolphins' repertoire of whistles and were used to label objects in the water such as toys or sargassum. Dolphins have demonstrated an uncanny knack for mimicking computer-generated whistles (Richards, Wolz and Herman 1984), and the goal with CHAT was to teach dolphins the "names" of these objects by inducing them to produce the correct whistle in context (e.g., they mimic the whistle for "sargassum" when playing with sargassum). So far, the success of this device for establishing two-way communication has been limited, but this still has profound implications for interstellar communication insofar as it suggests that the possibility of communication has a lower bound. In other words, once a species has developed symbolic communication systems, it appears that they cannot revert to presymbolic modes of communication or thought. Although we can attempt to make up for our inability to naturally mimic dolphin whistles by artificial means like CHAT, at the end of the day we are still indoctrinating the dolphins into the symbolic regime rather than learning the meaning of dolphinese.

This also raises a related but opposite issue that has always haunted research on interspecies communication: what if these instances of nonhuman primates and cetaceans ostensibly understanding symbolic communication systems are just examples of rigorous conditioning? After all, the nonhuman primates that learned basic addition, for instance, took a decade to do so, while human children can learn similar tasks after only a few examples. If interspecies communication really is indicative of animal understanding, then this lends credence to the hypothesis that the "basic computational biological prerequisites for human language" are already present in nonhuman primates, cetaceans, and perhaps other animals with higher intelligence. By exploring the difference between human languages and animal communication systems, we discover the minimal elements that a language designed for interstellar communication must possess to be intelligible. To the degree that intelligence implies the capacity for thought, and thought requires the capacity for language, METI assumes that extraterrestrial intelligences have language. Without these elements, communication with extraterrestrials may be impossible, or limited to the types of rudimentary exchanges typical of interspecies communication on Earth.

ENTROPY AND THE DOLPHIN

The foregoing was not meant to suggest that research on animal intelligence and communication systems is not relevant to METI, but was an attempt to delimit the relevant applications. Given that nonhuman animals do not have what we would call language but nevertheless possess complex communication systems, exploring the boundaries between language and communication is particularly relevant to the task of designing a message that will be recognized by extraterrestrials as an intelligent signal. A promising approach to this problem involves the analysis of animal communication systems from an information-theoretic perspective (McCowan, Hanser, and Doyle 1999). A remarkable feature of every human language is that the frequency of a word is inversely proportionate to its rank in the frequency table. In other words, the most frequent word occurs twice as often as the second most frequent word, three times as often as the third most frequent word, and so on (Zipf 1965). This relationship also holds for letters and phonemes. When the frequencies of these lexical units are graphed in logarithmic rank order it yields a slope of –1. The distribution of lexical units seen in all human languages appears to be necessary for a noncoded communication system to possess syntax (Ferrer, Cancho, and Solé 2003). Shallower slopes would indicate more diversity and less repetition in a system, ultimately culminating at a slope of 0, which would indicate an entirely random system.

Although Zipf slopes are good indicators of higher-level complexity in a signal, they don't give the whole picture. This is because they give equal weight to each point plotted outside of the context in which they occur, even though there will be significantly more data points for the most frequent unit than the least frequent unit of information in a communication system. Like Zipf's law, entropy is a universal feature of human languages (Montemurro and Zanette 2011), so quantifying the entropy of a communication system by analyzing the frequency of the units can help determine the complexity of a communication system and whether it qualifies as linguistic. Shannon entropy, which quantifies the information content of a communication system, can be measured at different levels, or "orders," with each increasing order corresponding to an greater degree of complexity in terms of the relationship between units of information in that system. In a system with zero-order entropy, for instance, each unit of information would be independent of the one that preceded it. It would, in other words, be entirely random. Zipf's law, by comparison, ranks lexical units by frequency of occurrence and thus examines a first-order entropic relationship (McCowan, Hanser, and Doyle 1999). An analysis of second-order Shannon entropy would evaluate the frequency of pairs of lexical units; third-order entropy would range over triplets of lexical units; and so on. Each increasing order provides a measure of the information content at that level, and the number of levels can be used to determine the complexity of a communication system. (Human languages appear to "bottom out" in terms of structure around the ninth entropic order; beyond this, the relationship between lexical units appears random.)

If dolphin whistles are considered as discrete units of meaning analogous to human words, the relationship between the frequencies of dolphin vocalizations yields a Zipf slope of –0.95, which is remarkably close to that of human language (Doyle et al. 2011). This suggests that dolphin communication has some higher-order structure (i.e., it may possess syntax). The case with Shannon entropies, however, is remarkably different. The first four entropic orders for English, Russian, and Arabic letters, for example, show a remarkable consistency across the languages, generating slopes of –0.500, –0.566, and –0.797, respectively. The zero-order entropy—which indicates sample diversity—of dolphin whistles (4.75) is similar to the distribution of letters in the three human languages (5.00, 4.75, and 5.00), but the entropic order of dolphin whistles fall well below the range of human languages in the first, second and third orders, yielding a slope of –1.334. A possible explanation for why dolphins demonstrated Zipf slopes similar to human languages but wildly different values for higher-order Shannon entropies is undersampling, which provides less statistical information over larger groupings of information units (McCowan, Hanser, and Doyle 1999). A more accurate assessment of entropy in dolphin communications, and the extent to which they can be considered an analog of natural human language, will depend on the collection of more robust samples of dolphin whistles.

If Shannon (1948) is right and the "fundamental problem with communication is reproducing at one point either exactly or approximately a message selected at another point"—a characterization that will certainly find sympathy among those designing extraterrestrial messages—then the ability to quantify the entropy of an interstellar communication system is critical to ensuring its intelligibility to the receiver. If we were to compose a message demonstrating zero-order entropy, for instance, there would be no way for the recipient to tell that it was intelligent in nature or whether it was receiving the message as intended by the sender. To this end, Laurance Doyle has suggested that Zipf's law and Shannon entropies could be used as a sort of SETI "intelligence filter" to analyze incoming signals from outer space to quantify their complexity (Doyle et al. 2011). Conversely, designers of messages intended for extraterrestrial intelligences can subject their own messages to this sort of intelligence filter. This could aid in its recognition by any extraterrestrial recipients and alert them to whether the message is linguistic in nature.

Following Doyle, let us consider two examples that are relevant to interstellar communication. An interstellar message must distinguish itself from "ordered" astrophysical phenomena such as pulsars for a receiver to recognize it as artificial. It is hardly surprising that the first pulsar to be discovered was mistaken as an extraterrestrial beacon, but can a pulsar be eliminated as a possible intelligent signal based on its information content alone? By analyzing 20,000 pulses collected from the Vela pulsar over half an hour (Johnson et al. 2001), Doyle and his colleagues were able to determine that the amplitude modulation of these pulses yielded a Zipf slope of -0.69 and an entropic slope of -0.28. To see the importance of applying an "intelligence filter" to interstellar communications, now consider the information content of the first message sent into space. The 1974 Arecibo message yields a Zipf slope of -1.64, which might indicate to an extraterrestrial recipient that the message is too redundant to be a complex communication system and might help the receiver determine that the binary code is meant to be arrayed as a bitmap. On the other hand, a plot of the message's entropy from the zero to the seventh order yielded an entropic slope of -0.03, which suggests a nearly random distribution of the binary digits within the message. This might cause an extraterrestrial recipient to miss the message entirely by dismissing it as random, meaningless noise. The astute observer will notice something strange about this comparison, however. From an information-theoretic perspective, a rapidly rotating star seems more likely to be a communication system than a painstakingly designed intelligent message. The reason for this is that "information theory is not so much about what you *are* saying as what you *could* say" (Shannon and Weaver 1949). Any analysis relying solely on information theory won't capture the meaning of a signal, but it will be able to quantify the ability of a system to handle communications governed by rules of arbitrary complexity. As Doyle and his colleagues note, if we were to receive a signal that began with the Fibonacci series (1, 1, 2, 3, 5, 8, 13, 21 ...) we would immediately recognize its significance, but from an information-theoretic perspective this signal would appear to be random. Another important point is that information theory can be used to distinguish astrophysical phenomena from intelligent signals by looking at the variation in the structural dependencies of communication systems (or lack thereof). Natural languages demonstrate a steady decline in structural dependencies between units of information as the entropic orders increase, whereas the dependencies between units of information in the Vela pulsar "communication system" stayed the same as the entropic orders increased. This suggests that it may make sense to begin an interstellar broadcast with a natural language text, rather than a magic number (e.g., pi or the Fibonacci sequence) or a logical metalanguage, in order to distinguish the signal as linguistic in nature to any extraterrestrials that might be using an information-theoretic intelligence filter of their own.

4

COSMIC COMPUTERS AND INTERSTELLAR CATS

In 2036, any inhabitants of the HIP 4872 solar system in Cassiopeia will receive a strange visitor. Her name is Ella, and she enjoys playing Atlantic City blackjack, telling jokes, predicting fortunes, and reciting poems. These hobbies are not all that unusual for a human, but Ella isn't exactly human. She's a chatbot: a natural language processing algorithm that can reproduce human speech by analyzing patterns in large collections of text. Ella's software was included as part of the 2003 Cosmic Call message, and it remains the first and only artificial intelligence sent into interstellar space. Shortly before transmission, Ella, who was created by the software company EllaZ systems, won first place in the Loebner Prize Contest, an annual Turing test competition in which judges try to distinguish humans from chatbots by holding textual conversations (Copple 2008). By today's standards, when many of us have a phone with far more advanced language-processing algorithms in our pocket, Ella comes off as a very crude approximation of intelligence, but at the time the program was considered one of the "most human computers" in the world. We needn't worry ourselves too much that our first AI ambassador to the stars might come off as an incoherent gambling addict, however. Without a primer in the syntax of

Visual Basic.NET, the programming language used to write Ella's software, there's a strong chance that extraterrestrials wouldn't be able to interface with the chatbot; but Ella's English language corpus, included with the software, could be a valuable reference material for them. Despite the shortcomings of "astrobot Ella," its transmission to Cassiopeia was a landmark event that pointed to a promising future for the use of AI in interstellar communication.

The notion of sending artificial intelligences or digital human avatars as extraterrestrial envoys has a long history in popular science fiction, so it's hardly a surprise that it was one of the first ideas considered for an extraterrestrial message by the pioneers of modern METI. During a particularly lively discussion at the first Soviet–American conference on extraterrestrial intelligence in 1971 at Byurakan, the attendees debated how an extraterrestrial intelligence might interpret a message. Against the idea of communicating using artificial languages such as Freudenthal's *lingua cosmica*, which requires a method for teaching how to decode the language, one of the Soviet delegates suggested sending self-evident information—such as a drawing of a cat. The cat would be drawn in the three-dimensional space of Euclidean geometry, where each of the coordinates is derived from parameters of the signal itself (e.g., frequency for the y-axis, time for the x-axis, and signal intensity for the z-axis). This process could be repeated for a virtually unlimited range of ideas. Furthermore, Kuznetsov argued that by repeating the same picture at multiple frequencies, these information-laden signals could also serve as a sort of beacon to attract an extraterrestrial's attention in the first place. Marvin Minsky, however, had a different idea. "Instead of sending a very difficult-to-decode educational message of the kind that Freudenthal describes, and instead of sending a picture of a cat, there is one area in which we can send the cat itself," Minksy said. "Briefly, the idea is that we can transmit computers" (Sagan 1973).

Minsky pulled this idea straight from science fiction. As an example of his proposal he cited *A for Andromeda*, a television series written by the cosmologist Fred Hoyle in which Earth detects an extraterrestrial signal that contains instructions for building a computer that then relays instructions for creating a biological organism named Andromeda. Despite its fictional origins, Minsky's suggestion was still pragmatic. If an extraterrestrial planet had a different refractive index than that of Earth, transmitting a picture using Kuznetzov's coordinate system would result in a distortion of the picture from the extraterrestrial's point of view. Although the topology of the signal would be retained, a distorted picture would undermine the image's usefulness, especially if the picture was attempting to convey a scientific concept. This suggests that an ideal message would be entirely topological (i.e., only containing properties that are immune to distortion). It was Minksy's insight that the computer is an "absolute topological device" and could thus be considered the ideal content of an interstellar message. To teach an extraterrestrial how to build a software program, or better yet, a computer running a desired software program, Minsky suggested a series of instructional pictures arranged like Drake's prime number bitmaps that would outline the design of the computer and software. Furthermore, Boolean logic diagrams could be used to eliminate any uncertainties about the contents of the signal. The computer itself could be used to run software similar to Freudenthal's *lingua cosmica*, which would help eliminate decoding problems in the Lincos system through feedback. Furthermore, the computer program could be designed to interact with the extraterrestrial intelligence so that it could, for instance, employ natural language processing algorithms to learn the language of its hosts.

LANGUAGE CORPORA AND ETAI

When Chomsky first published his theory of generative grammar, it was partly in reaction to the dominant trend in linguistics in which language was effectively reduced to Markov processes. These describe regularities in a system in which the next unit in a sequence is probabilistically determined by the values of the units that preceded it. Chomsky's famous counterpoint to the use of statistical models for language syntax is the sentence "Colorless green ideas sleep furiously," which is grammatically correct, but the probability of its production in English is statistically negligible (Chomsky 1957). Thus, linguists went off in pursuit of the deeper rules that allow for such nonsensical yet grammatically valid sentences, while the computer scientists continued to explore natural language patterns with statistics. The significance of this division was not lost on Minsky. "I claim that most information that people have that is important is not facts, but processes," Minsky said at the Byurakan conference. "In particular, the process by which you parse a language and understand a grammar is much more important than the grammar itself" (Sagan 1973). What Minsky is essentially describing is natural language processing (NLP), a branch of machine learning that analyzes human language so that a machine can "understand" and respond in natural language. In the early days of NLP, programmers would code explicit rules into a computer to teach it how to analyze natural language (Winograd 1972). The issue with this, however, was that humans were limited in their ability to explicitly capture the manifold ways that natural language is used in practice. The so-called statistical revolution in natural language processing that began in the early 1980s overcame these early limitations by feeding computers large samples of text, known as corpora, and allowing algorithms to tease out the multitude of nuanced rules at play in the use of natural language (Johnson 2009). Today, statistical analysis still dominates NLP algorithms, although their effectiveness has been augmented by a class of machine learning algorithms loosely based on the structure of the human brain called artificial neural networks. Unlike the statistical NLP models of yore, deep learning algorithms can categorize the data they use to perform a specific task, such as parsing the grammar of a sentence or extracting semantic content from an utterance (Socher, Perelygin, et al. 2013; Socher, Bauer, et al. 2013). With respect to deep learning NLP algorithms, however, the fundamental process is still the same insofar as it involves classifying parts of speech and doing statistical analysis on natural language corpora.

When someone refers to "artificial intelligence" in everyday conversation, he or she is usually referring to machine learning, a narrow form of AI that is a far cry from the artificial general intelligence (AGI) of science fiction. AGI is the ultimate goal of machine learning research, but achieving this goal turned out to be far more difficult than the pioneers of AI imagined (Wang, Goertzel, and Franklin 2008). Given that any extant extraterrestrial civilizations are likely more advanced than our own, it is reasonable to presume that an extraterrestrial intelligence has developed its own AGI (Shklovskii and Sagan 1966). Indeed, there is good reason to presume that *most* extraterrestrials in the universe are "postbiological" artificial intelligences (Bainbridge 2013; Dick 2006; Sandberg, Armstrong, and Cirkovic 2017).

Operating on this assumption creates space for interstellar messages designed according to principles of natural language processing on Earth, which come with numerous advantages as far as detection and decoding are concerned.

As with natural language processing on Earth, an interstellar message destined for extraterrestrial artificial intelligence (ETAI) would begin with selecting a natural language corpus (Atwell and Elliott 2001, 2002). The content of this corpus would be transmitted in binary, with strings of fixed length representing each of the characters in the corpus according to some standard (Unicode, ASCII, etc.). Yet the question of which corpus to select immediately poses a problem. To this end, John Elliott (2011) detailed the design of a human language chorus corpus (HuLCC) that would capture all aspects of typology across every human language. Even if only a single terrestrial language is encoded in an interstellar transmission, HuLCC would assist in distilling the universal mechanics of terrestrial languages into "grammatical, cognitive, and structural design templates for rationalizing message construction to expedite the recipient's ability to detect and decipher message content." Per Elliott's design, this human language corpus would actually be a collection of subcorpora in every terrestrial language, ranging across 20,000 words of the same text.

There isn't a single text that has been translated into every extant language. Even the Bible, the mostly widely translated text in history, has only been translated into about 10 percent of the roughly 7,000 living languages. Thus, the creation of HuLCC will almost certainly involve a robust translation effort. Although religious texts such as the Bible or Quran appear to be a good starting point, they violate Elliott's maxim that the core text ought to be written in the contemporary vernacular, so perhaps other widely translated fictional texts, such as Hans Christen Andersen's *Fairy Tales* or Jules Verne's *Twenty Thousand Leagues Under the Sea* would be adequate alternatives. The adequate size of a corpus is a contested subject among linguists, although the common wisdom is that when it comes to corpora, the bigger the better. Interstellar messages are subject to energy constraints that limit the amount of information that can be sent. Thus, a more relevant consideration is the minimum size of a corpus needed to produce reliable results. In this respect, corpora above 14,000 words have been shown to produce reliable machine translations, but about 20,000 words is the necessary threshold to distinguish the text as linguistic material rather than a random collection of lexical units (Elliott 2011b). As seen in various quantitative analyses of animal and human communication systems (Lee, Jonathan, and Ziman 2010; Doyle et al. 2011; McCowan et al. 1999), the distribution of lexical units in natural language corpora have distinctive signatures (e.g., Zipf's slope and predictable variation across increasing entropic orders). Although power laws similar to Zipf's slope can be identified in natural phenomena such as the molecular distribution in the DNA of yeast (Jensen 1998), the distribution of these units usually occupies a far greater range and demonstrate long repeating patterns far from one another in the "corpus" (Elliott, Atwell, and Whyte 2000a,b). Moreover, Elliott notes that in any given language, the actual combination of symbols in pairs and triplets is far lower than the actual number of possible combinations: Only about 50 percent of all possible two-letter combinations are used and only 20 percent of all possible three-letter combinations are used, mostly owing to restrictions on sound production. Elliott contrasts this with randomly generated text, which can use all possible combinations within an equivalent of about 20,000 words, which serves as a conservative lower bound on the minimum size of language corpora for interstellar communication. Taken together, these characteristics may identify the signal as linguistic in nature to the extraterrestrial intelligence.

An important issue related to the selection of a natural language corpus for interstellar communication is the selection of an annotation mechanism for the linguistic data. Although deep learning algorithms are becoming increasingly adept at receiving large unmarked linguistic datasets and analyzing the contents of these data without human supervision, many NLP algorithms still depend on well-annotated data to learn a language efficiently (Stubbs and Pustejovsky 2012). Language corpus annotations are metadata that describe linguistic features relevant to understanding the text, such as parts of speech or the sentiment of a collection of words. The value of this metadata for deciphering the structure of a natural language interstellar message is readily seen by considering the difficulties faced by archaeologists attempting to understand lost languages on Earth. In the case of lost languages like Linear B, which was first discovered on Crete in the late-nineteenth century, one of the most significant barriers to understanding its contents was simply parsing lexical units (Chadwick 1990). Consider the case of extraterrestrial intelligences receiving an interstellar signal of unknown origin. How are they to determine the constitution of individual characters in this massive stream of binary information? As we saw above, structural elements of natural language allow for an unsupervised natural language algorithm to identify characters through statistical analysis. For example, if the natural language corpus was coded according to ASCII standards, where each character is represented with a unique 8-bit

ID, an extraterrestrial intelligence would be able to analyze the corpus for the entropy associated with different bit-lengths. This analysis would reveal a substantial decrease in entropy for 8-bit blocks of the signal, suggesting that each individual character corresponds to a byte of information. If a unique identifier for a "space" is included in the message as well, a similar technique could be used to identify individual words by analyzing the message for space characters. Once this has been accomplished, an analysis of the distribution of the words within the corpus would reveal an organized structure corresponding to Zipf's law (Elliott, Atwell, and Whyte 2000a,b). Although all of this could be accomplished without any textual annotation, the inclusion of this metadata would help ensure the correct interpretation of the corpus by the ETI. As we will see in the next chapter, the Lincos metalanguage developed by Alexander Ollongren would be a natural choice for the annotation of an interstellar corpus. Its simple syntax based on the lambda calculus makes it easily readable, and its self-interpreting design is informed by the calculus of constructive inductions so that Lincos statements can be checked for correctness within the system itself. Thus, Ollongren's Lincos can serve as a sort of algorithmic Rosetta Stone to help guarantee the correct interpretation of a corpus-based interstellar message.

COSMIC OS

Although Ollongren's Lincos may have marked the algorithmic turn in language design for communication with extraterrestrials, he wasn't the first to apply computer science to the problem of interstellar communication. In fact, he wasn't even the first to apply principles of software development to METI. This can properly be credited to Paul Fitzpatrick, a computer scientist at MIT who developed Cosmic OS, an operating system modeled on the Scheme programming language. Originally developed in late 2003 before being "rebooted" in 2014, Fitzpatrick's operating system is meant to function as a standalone computer program that is capable of being understood by an extraterrestrial intelligence and to convey a "significant portion of the human world view" in the process.

Fitzpatrick wrote Cosmic OS as a derivative of the 1970s-era programming language Scheme. This functional programming language was originally developed at MIT's AI lab by Guy Steele and Gerald Sussman as a dialect of Lisp. Except for Fortran, Lisp is the oldest high-level programming language still in use today and is notable for its reliance on the lambda calculus. Outlined in a series of memos published between 1975 and 1980, known as the Lambda Papers, Scheme was born from the realization "that the lambda calculus—a small, simple formalism—could serve as the core of a powerful and expressive programming language" (Sussman and Steele 1998a,b). Although both Scheme and Lisp have a common basis in the lambda calculus, Scheme offered a suite of improvements over Lisp, particularly with the introduction of lexical scope. In a computer program, the scope of a name binding—the association of a name with a variable—determines where in that program the binding is valid.

With Cosmic OS, the goal is to use this language to develop a multiuser dungeon (MUD) that could be run by extraterrestrials in order to learn more about Earth and the people who inhabit it. MUDs are essentially text-based virtual worlds that are somewhere between a role-playing game and an interactive fiction. For Fitzpatrick, this would assume the role occupied by the "morality plays" found in Freudenthal's Lincos (for a more detailed discussion of the Lincos morality plays, see appendix C), but with a better schema for correction and the ability to transmit more information.

An interstellar transmission including Cosmic OS would have a "strong backbone of actual executable code," and the execution of this code would allow the ETI to interpret the information conveyed via a MUD. According to Fitzpatrick, this has two main advantages. In the first place, recipients can look at the code in detail to understand how it works or treat the system as a black box and try to learn how it operates by comparing input and output. Moreover, by understanding the details of the code, the recipients could alter it so that they could derive more information from the code than was included in the original message, as well as craft a response using similar principles. Of course, Cosmic OS assumes that the message's recipient has a computer that can execute the code, but this was a reasonable assumption adopted by Fitzpatrick after considering the amount of technological sophistication a receiver would have to have to pick up the message in the first place. In its final form, the messages sent via Cosmic OS would consist of just five characters: the binary digits 1 and 0, open and closed brackets, and a semicolon to delineate sentences. The simplicity of its lexicon is a major boon for the project insofar as it requires minimal bandwidth and would be easily interpretable by a computer on the receiving end. The initial message would begin by teaching the recipient how to count, introduce operators such as equality, greater than, less than, and other logical notations such as "not" and "and."

By the second iteration of Cosmic OS, Fitzpatrick had stripped the language down to a simpler functional programming language consisting of eight symbols, each of which has a corresponding number from 0 to 7 and a functional meaning. For example, the symbol "." corresponds to the number "1" and its meaning is a function that adds one to its argument. The numbers 3 and 4 respectively correspond to the symbols for open and closed brackets and are used to mark the beginnings and ends of expressions. The third iteration simplified things even further, reducing the language's lexicon to just five symbols: a period and colon corresponding to the binary digits 1 and 0, open and closed brackets to mark the beginnings and ends of expressions, and a semicolon which marks the end of a sentence. In this notation, numbers are encoded as binary digits within parentheses, so that (::::.) would equate to 14, sets of numbers are included between parentheses and constitute an expression, and these expressions can be nested. By this iteration, Fitzpatrick had also developed a suite of twenty-nine math-based lessons, albeit in condensed, skeletal forms, which used this notation to teach everything from counting to simple mutable structures. Importantly, these lessons introduce lambda notation, which makes the program quite difficult for a human to read, but is nevertheless a highly expressive notation, in addition to a relatively sophisticated MUD that is essentially a collection of statements to create a world that consists of things that act like locations and people. For example, in this iteration of the MUD, a robot character travels through a variety of rooms and spaces, such as a hallway, kitchen, front lawn, and bedroom. At this point, the program was sophisticated enough to "introduce a language to talk about what is going on in the simulated world, and start to move away from [the] detailed mechanism," but as Fitzpatrick notes, isn't developed enough that it would be very understandable to the recipient just yet (Fitzpatrick 2014).

The idea behind Cosmic OS is that by beginning with simple math, it is possible to construct a programming language that can simulate an interactive virtual environment for an extraterrestrial intelligence. Such a rich environment would in principle allow the extraterrestrial to manipulate the program to get a better idea of the social and behavioral properties of the Earthlings who sent the message. Beyond this, the recipient would also be able to look at the source code for the simulations, which would provide a rich text for understanding the logical imperatives of the simulated characters and insight into human cognition. Cosmic OS represents a unique approach to interstellar message construction, but several difficulties in the language still require attention. For instance, it is necessary to find the best way to encode the message as a physical signal, as well as test its robustness against interference. Fitzpatrick is still in the process of developing simulations and alternative introductions to the language, the idea being that offering alternative explanations of the message increases the chance that the recipient will understand it. To this end, Fitzpatrick is exploring the use of two-dimensional images of logic circuits that could be included in the message as alternative explanations of the operating system's kernel.

DNA AS EXECUTABLE CODE

If aliens were to visit Earth and learn about its inhabitants, would they be surprised by such a wide variety of life all sharing a common underlying genetic code, or would this be all too familiar? There is probable cause to assume that the structure of genetic material is the same throughout the universe and that, while this is liable to give rise to life forms not found on Earth, the variety of species is fundamentally limited by the constraints built into the genetic mechanism. If we may briefly return to the topic of sending a cat through interstellar space, could it be that the best option is to send the cat's genetic code? On Earth we have only sequenced the genomes of a small percentage of living organisms and have only recently completed the human genome. We have successfully cloned several animals, but technical and ethical roadblocks prevent scientists from doing the same with humans, although the technical barriers to this pursuit are rapidly eroding (Liu et al. 2018). If an extraterrestrial civilization isn't burdened with ethical dilemmas about cloning, however, sending the genetic code for humans and other species may be the most effective way to teach them about our biology.

References to our genetic makeup have been a feature of interstellar messages from the very beginning. Although the first genes wouldn't be sequenced for another three years, the 1974 Arecibo message included a rudimentary bitmap of DNA's helical structure. A quarter of a century later, both Cosmic Call messages included symbols for each of DNA's four nucleotides. To date, however, only a single interstellar transmission has encoded any genetic information. To commemorate the thirty-fifth anniversary of the Arecibo message, the artist Joe Davis travelled to Puerto Rico to broadcast the genetic sequence for the large subunit of the ribulose-1,5-bisphosphate carboxylase oxygenase (RuBisCO) molecule. RuBisCO is the most abundant protein on Earth and plays a major role in converting atmospheric carbon dioxide into energy-rich molecules for plants. To encode this genetic information in a signal, Davis first considered representing each of the 1,434 nucleotides with a two-bit ID (C=00, T=01, A=10, G=11) to create a 2,868-bit sequence representing the RuBisCO molecule. The problem with this, of course, is that there isn't enough information to use analysis techniques such as those described above by Elliott. Thus, any ETI that received this message would have no way to determine the coding schema used to create the message, which would essentially be an unintelligible mess of data.

For better or worse, it is unlikely that any extraterrestrials will ever receive, much less understand, Davis's message. None of the stars selected by Davis have been confirmed to host planets, and two of the target stars aren't likely able to support life even if they do. GJ 83.1 is a flare star, a type of dwarf known for periodic bursts of intense radiation and Teegarden's star is a red dwarf, a type of star that is widely believed to be too cool to support life unless the planet was so close to the star that it would become tidally locked, meaning that half of the planet would be in perpetual night. Even if there are intelligent inhabitants around any of the three "RuBisCO stars," the odds that they would be able to interpret Davis's message is quite low, given the lack of context or redundancy to correct for message corruption during transit. Davis was the first to admit that his interstellar message was meant more for his fellow passengers on spaceship Earth than extraterrestrials, but this stunt points the way to a promising future for METI. In the last few decades, biologists have sequenced the genomes for thousands of species, including humans. These are effectively the "blueprints" for the species, but we are only just beginning to learn how to read the code. A sufficiently advanced extraterrestrial intelligence may have advanced genetic engineering to the point where genomes are the equivalent to an executable computer program, which would allow them to artificially recreate a human and other terrestrial species in their own labs. This assumes that extraterrestrials are made of the same genetic "stuff" as life on Earth, but as we will see later this is not as large of an assumption as it first seems. If an interstellar DNA message is accompanied with an instruction manual that explains its contents and the environmental parameters necessary for the organism's survival, Minsky's dream of sending a cat across the cosmos could very well become a reality.

5

IS THERE A LANGUAGE OF THE UNIVERSE?

The primary difficulty when it comes to METI is finding common ground between ourselves and other intelligent entities about which we can know nothing with absolute certainty. This common ground would be the basis for a truly universal language that could be understood by any intelligence, whether in the Milky Way, Andromeda, or beyond the cosmic horizon. To the best of our knowledge, the laws of physics are the same throughout the universe, which suggests that the facts of science may serve as an adequate basis for mutual understanding between humans and an extraterrestrial intelligence.

The first symbolic system designed explicitly for interstellar communication and based on scientific facts was the Astraglossa, first proposed by the experimental zoologist Lancelot Hogben in the late 1950s. Although our "extraterrestrial neighbors" are unlikely to understand any of our natural languages, Hogben argued that this is a situation that is not entirely unfamiliar to Earthlings. There are roughly seven thousand living languages and dialects spoken on Earth, a linguistic diversity that frequently impedes communication. Nevertheless, some aspects of language are widely shared across linguistic groups, and it is these "common fields of semantic reference" that Hogben suggested could form the foundation for an interstellar message. The most obvious choice, in this respect, are the natural numbers, which are found in every linguistic group on Earth. (Except Pirahã, the language of the isolated Pirahã people in the Brazilian Amazon, which doesn't include numbers; Everett 2005.) The second is celestial events, such as the phases of the moon or the position of the planets in the sky. Thus, Hogben suggested that an interstellar message ought to be designed such that "numbers will initially be our common idiom of reciprocal recognition ... [and] and astronomy will be the topic of our first factual conversations" (Hogben 1952).

Hogben framed his Astraglossa project as an esoteric exercise in autodidacticism. Given the unique constraints involved with interstellar communication, the thought experiment posed by the Astraglossa can teach us a lot about how we teach ourselves. In fact, Hogben saw his cosmic language as an attempt to teach extraterrestrials how to teach *us*. Unlike the "shipwrecked mariner [who] learns the lingo of the North Caroline Islands" by pointing to objects to learn the basics of the language, Hogben realized that the astronomer attempting to communicate with extraterrestrials must first devise a system that shows how to point at things. Hogben likened the interpretive dilemma faced by an extraterrestrial recipient of a message coded in Astraglossa to that faced by the first Europeans to encounter the Mayan civilization in the sixteenth century. Unlike the European decimal system, the Mayans used a vigesimal numeral system comprised of stacked dots and dashes. Despite the foreignness of this system, however, it shared some fundamental syntactical components that enabled interpretation: iteration, rank order, and gap. As Hogben noted, these syntactical elements can be translated into a one-dimensional (temporal) medium of communication such as a radio signal and can reasonably be assumed to be understood regardless of the specific numeral system used by an extraterrestrial intelligence. Thus, the Astraglossa notational system uses "strokes and dots," analogous to long and short radio pulses, to communicate basic math principles. Hogben's prototype interstellar message begins by introducing the concept of identity and basic arithmetic operators and builds the system from these basic components to the point where they can be used to describe celestial events and other complex topics. To communicate these ideas, Hogben introduces the concept of "radioglyphs," or groupings of individual signals that are "recognizable as a gestalt."

To discuss celestial events with an extraterrestrial, it would be necessary to know a good deal about how the cosmic environment would appear from the extraterrestrial's perspective. At the time the Astraglossa was published, astronomers had yet to discover any exoplanets orbiting other stars, so Hogben posited communication with Martians as a hypothetical example of his system in practice. This would first require terrestrial astronomers to construct a reference calendar for an event such as the sunrise or sunset on Mars relative to some fixed point. From this fixed point of reference, it would be possible to discuss celestial events, such as the conjunction of Earth and Venus, from the point of view of the Martians. At this point, it is conceivable to use these reference points to construct increasingly complex ideas, and eventually Earthlings and the Martians could communicate things about their own planets, like the composition of the planet's crust, its mass, or diameter. The same basic idea would apply to any extraterrestrials that may exist on the hundreds of exoplanets that have been discovered to date. Astronomers can tell a great deal about many of these exoplanets, such as their orbital inclination, mass, and with the next generation of space-based planet hunting telescopes, perhaps even details like the composition of the planet's atmosphere.

After discussing celestial events, Hogben suggested that interstellar messages could progress to the teaching of vocatives, such as "we" or "you." In this way, Hogben managed to bolster his syntax so that it was capable of conveying not only queries, affirmations, and negations, but also doubt, conditionals, and causal relationships. At this point, he argued, it would be possible in principle to challenge the extraterrestrials to a game of cosmic chess to "divert some of the deplorable combativeness of our own species by recording interplanetary tournaments to keep the international news out of the headlines." Humor aside, Hogben realized that his language limited the topics of our cosmic communications to gamesmanship and the natural sciences. In this respect, the Astraglossa was incredibly impoverished. Presumably, an extraterrestrial recipient of a message coded with the Astraglossa would want to know something about the individual experience of the intelligence that crafted the message. To this end, Hogben rounded out his radio lexicon with a radioglyph meant to convey the notion of "ego." By introducing this concept of individuality, Hogben made it possible to communicate several ideas about our physical characteristics, the biological mechanisms of the body, and what it means to die.

For all its ingenuity, the Astraglossa was burdened with a fatal flaw: It assumed reciprocity. Although a suite of exoplanet-hunting telescopes have revealed that there are likely far more planets than stars in our galaxy (von Braun and Boyajian 2017), SETI has yet to turn up any indications that those exoplanets host intelligent life. As the search for extraterrestrial intelligence extends to ever-farther reaches of the galaxy, it appears that if we *do* hear from an extraterrestiral, that message will likely have been traveling across the cosmos for hundreds, if not thousands, of years. While this doesn't necessarily preclude a response, the timescales involved are so long that it makes the Astraglossa a wildly inefficient system, since it can only communicate a few basic mathematical and scientific facts without reciprocity. So, while communicating mathematical concepts and astrophysical knowledge appears to be a good basis for an interstellar message, its usefulness depends largely on the degree to which the message is self-interpreting. Ideally, the extraterrestrial recipient ought to be able to decipher it without sending us a reply and waiting for feedback.

In 1990, the mathematician Carl DeVito and linguist Richard Oehrle outlined a proposal for an interstellar communication system "based on the fundamental facts of science." The message would be self-contained and would do away with the need to communicate much about the language itself before beginning to communicate interesting or useful information, thereby conserving two of the most precious commodities in interstellar communication: time and bandwidth. This scientific language assumes that extraterrestrials will have a strong grasp of scientific principles on the basis that they must be able to manipulate electromagnetic radiation to receive the message in the first place. A message coded using this symbolic system would begin with presumably shared scientific knowledge to communicate notation for numbers, logical operators, chemical elements, and our basic units of measurement. After this rudimentary lesson in terrestrial notation, the system could be used to communicate precise and interesting information. For example, the fact that we live on a planet is rather uninteresting, but information about the mass, atmosphere, or orbital velocity of this planet would presumably be of great interest to an extraterrestrial recipient. DeVito and Oehrle envisioned this message being transmitted in a series of stages, in which extraterrestrials would first receive an outline of numeral and logical notation, before progressing to chemistry, physics, and so on. While it would be relatively straightforward to continue discussing increasingly complex topics *within* a domain, be it mathematics or chemistry, the real challenge would be to move between domains (i.e., from basic mathematics to chemistry). To do this, scientific facts that can be considered universal, such as an element's atomic weight, would serve as "links" to communicate facts that might be unique to Earth, such as the boiling point of water.

Yet how can we be sure that a language based on scientific facts will be adequate to establish communication with an extraterrestrial intelligence? After all, our sciences looked much different only a century ago, when the theory of general relativity was still just a gleam in Einstein's eye and the structure of the atom was conceived in terms of Bohr's model. If Earth had broadcast a message consisting of scientific facts to extraterrestrials at the turn of the twentieth century, its recipients might choose to ignore a message from such a scientifically ignorant civilization. Moreover, the timescales involved in reciprocal interstellar communication will most likely be measured in centuries, if not millennia, so we must consider how much our scientific knowledge will progress between the time we send a message and receive a reply. This suggests we may want to look for something even more fundamental than scientific facts as the basis of an interstellar message.

WILL EXTRATERRESTRIALS UNDERSTAND OUR MATH?

If science is the process of systematically describing the universe, then the language of science is, by analogy, the language of the universe. This is an observation that is as old as modern science itself and was perhaps most succinctly summarized by Galileo in *The Assayer*, his foundational treatise on the scientific method. "This grand book, the universe, stands continually open to our gaze," Galileo wrote. "But the book cannot be understood unless one first learns to comprehend the language and read the letters in which it is composed. It is written in the language of mathematics, and its characters are triangles, circles, and other geometric figures without which it is humanly impossible to understand a single word of it." The notion that mathematics is language of the universe has haunted the development of science ever since Galileo's pronouncement and was the source of bewilderment for history's most eminent scientific thinkers. The conundrum was eloquently described by Albert Einstein (1922), who wondered "How is it possible that mathematics, a product of human thought that is independent of existence, fits so excellently the objects of physical reality?" This question is, ultimately, about the ontological status of mathematics, and it has profound implications for designing messages for extraterrestrial intelligences. If mathematics exists independently of the human mind, then it is an ideal candidate for an interstellar message and one that would be immediately intelligible to an extraterrestrial recipient. Yet if mathematics is a human invention, is there any reason to suppose that it would be intelligible to an extraterrestrial?

The "unreasonable effectiveness" of mathematics when it comes to describing the physical universe has long been taken as evidence of the existence of abstract mathematical objects, a perspective known as mathematical Platonism. According to the Platonists, mathematics would exist regardless of whether there were any minds—human or otherwise—around to perceive it, because mathematics is literally a part of the universe and endows it with a rational structure. On this view, mathematics is something that is discovered, not invented. Almost every program for interstellar communication has tacitly adopted mathematical Platonism as an operating principle, which is hardly surprising given the disproportionate representation of mathematicians working on problems of interstellar communication and the strong support for mathematical Platonism throughout the discipline (Balaguer 1998). This philosophical position justifies mathematics as a starting point for interstellar messages because humans and extraterrestrials share a universe and are thus "reading the same book," even if the symbolic language we use to discuss that book may be quite different. From this perspective, METI can be largely characterized as an exercise in translation—an interstellar message isn't teaching an extraterrestrial about numbers, it's teaching them about numerals.

Consider DeVito and Oehrle's language based on the fundamental facts of science, which assumes that aliens are familiar with counting. As DeVito has argued, even though there are several philosophies of mathematics, "all of mathematics can be based on the notion of natural number," and as a consequence "all of our mathematics could, in principle, be communicated to any intelligent alien who understands these numbers" (DeVito 2011b). DeVito's approach to the philosophy of mathematics straddles the divide between the "extreme Platonists" and the formalists. The Platonists insist that all mathematical objects really exist independently of our mind and can thus be discovered just as an archaeologist discovers the ruins of an ancient city, whereas the formalists characterize mathematics more like a game that is played according to a set of well-defined rules. On this latter view, mathematical objects exist only through their applications; we might say there are three humans or seven oranges, but the numbers three and seven do not exist independently of our mental application of these concepts. Seen this way, mathematics is effectively an exercise in manipulating inherently meaningless symbols according to fixed rules. The consequence of this view is that it is conceivable that an extraterrestrial intelligence has no concept of the number three or seven, just as many ancient human societies had no concept of zero, or that their rules for manipulating mathematical objects are entirely different from our own. The formalist position does not necessarily preclude interstellar communication, however, but does suggest that mathematics may not be an adequate starting point for a message because the "rules of the game" must first be taught to the recipient. DeVito borrows from both mathematical philosophies insofar as he considers natural numbers to exist independently of the mind, even though the rest of mathematical objects may be a distinctly human idiosyncrasy. If it is the case that natural numbers serve as a base reality for a human fantasy world of mathematical objects, then DeVito acknowledges that mathematics "may be more reflective of our minds than we realize and may say more about human nature than it does about the real world" (DeVito 2011b). Indeed, there is good reason to suspect that our preference for whole numbers is a result of the constraints of the human mind; an extraterrestrial with a stronger short-term memory might not prefer natural numbers over, say, irrational numbers (Dehaene 2011).

One of the foremost conceptual problems with mathematical Platonism is how the human mind can come to know mathematics if the objects of the discipline exist as independent abstract entities on some metaphysical plane. Another serious challenge is that there is no empirical basis for the claim that mathematics is part of the structure of the physical universe. Like claims about the existence of God, it's a proposition that can't be scientifically tested. So, although mathematics has proven to be unreasonably useful in the natural sciences, the Platonists' explanation for why this is the case is unscientific. Yet rejecting mathematical Platonism doesn't get us any closer to understanding what mathematics actually *is*, which is of central importance to the design of interstellar messages.

Around the turn of the twentieth century, formalism and intuitionism emerged as leading critiques of mathematical Platonism. Formalism rejected the independent existence of mathematical objects and considered mathematics akin to a game in which meaningless signs are manipulated according to a set rules or formalisms. Although the formalist program was ultimately dismantled by Gödel (1992), who proved that any sufficiently developed formal system can produce a statement that can't be proven within that system, formalism laid the groundwork for digital computing and served as the basis for the first programming languages (McCarthy 1963). In terms of interstellar communication, Alexander Ollongren's second-generation *lingua cosmica* is based on the formalisms of the lambda calculus and calculus of constructive inductions, which are used as a metalanguage for the interpretation of a core text. Such formalisms are an attractive candidate for interstellar messages because the terms can be blindly manipulated without understanding their significance so long as the extraterrestrial understands the principles of logic that serve as the system's foundations. Intuitionism, on the other hand, also rejects the ontological necessity of abstract mathematical objects and posits that mathematics is a "languageless" creation of the human mind (Brouwer 1981). On this view, mathematics is a pure mental construction that proceeds from our experience of time. To the extent that symbolic systems are used in mathematics this is merely to convey these mental constructions to another mind. An important implication of Brouwer's intuitionism is the notion that a statement is true only by virtue of having a proof, which is used to construct the mathematical object in question. This intuitionistic logic led the way for algorithmic proof-checkers that can create proofs that are far too complicated to ever be intelligible to a human mind and is also at the core of Ollongren's *lingua*, which allows the system to be self-interpreting.

Although mathematical formalism and intuitionism have both played a significant role in the computing revolution, they fall short in their ability to explain what mathematics is and why it is so effective at describing the physical world. It wasn't until the advent of magnetic resonance imaging (MRI) in the 1970s that any truly scientific progress could be made toward answering this question. Armed with unprecedented access to the neural structures that underlie our experience of mind, cognitive neuroscientists began laying the groundwork for what would eventually become a theory of embodied mathematics (Lakoff and Núñez 2000). This revolution in brain imaging technologies led to an unambiguous conclusion: Our cognition is shaped by our bodily experiences. The very structure of our brain is optimized to allow our bodies to effectively engage with the world around us, which suggests that our embodied mind is the bridge between mathematics and the physical world.

Mathematical ability is, to a very limited degree, innate. Human infants, along with many other animals including rats and chimpanzees, have demonstrated the capacity for numerosity, or the ability to distinguish between the magnitudes of groupings of objects that haven't been counted (Dehaene 2011). Yet the capacity for numerosity is a far cry from the ability to do calculus or trigonometry, talents that adult humans alone possess. The question of how human infants make the jump from these innate mathematical abilities that are shared with other animal species to the complex mathematics is a scientific question that can be explored by studying the neural and cognitive basis of mathematical ideas. Núñez and Lakoff (2000) argue that mathematical ideas are heavily dependent on conceptual metaphor, which humans use to think about abstract ideas in concrete terms. Technically speaking, conceptual metaphor is the cognitive mechanism that implements a "grounded, inference-preserving cross-domain mapping" that allows us to "use the inferential structure of one conceptual domain (say, geometry) to reason about another (say, arithmetic)."

According to Lakoff and Núñez, humans have an innate ability to form metaphors and mathematics is grounded in reality through four primary conceptual metaphors that are borrowed from everyday experience of an infant, namely: object collection, object construction, measuring sticks, and motion along a line. Each of these four grounding metaphors can be derived from the innate arithmetic that is demonstrated by several species of animals and newborn infants in combination with the directly lived experience of the infant. Even though their arithmetical abilities are capped at around three or four objects, these objects can be grouped in collections, combined to form a new object, physically segmented, and moved through space. These experiences are mapped onto the domain of number that is structured by the infant's innate ability to subitize (the ability to recognize a small number of objects without counting them individually). Moreover, each of these four grounding metaphors can be conflated (for example, object construction also involves object collection). This results in neural links created by the coactivation of the areas of the brain that characterize those direct experiences and creates a cognitive isomorphic structure that is "independent of numbers themselves and lends stability to arithmetic." When the four grounding metaphors are mapped to the domain of number, they provide the basis for the laws of arithmetic, which can then be combined to form ever-more abstract and complex mathematical ideas. For example, the ability to add a physical item to another group of physical items or to take an item away from a collection metaphorically captures the mathematical ideas of associativity and the ability to construct an object, say, out of Lego bricks, provides the metaphorical basis for the concept of fractions, and so on. It's this metaphorical blending among our experiences of collections, the structure of objects, the manipulation of physical segments, and the experience of motion, according to Lakoff and Núñez, that is "the basis of the link between arithmetic and the world as we experience and function in it." Indeed, "it forms the basis of an explanation for why mathematics 'works' in the world."

The program of embodied mathematics described by Lakoff and Núñez is similar to the cognitive revolution in linguistics that established a link between the faculty of language and its physical implementation in the brain. The grammatical idiosyncrasies of natural languages are largely the product of historical circumstance, but the capacity for language as such is innate. Similarly, the form our mathematics takes is not arbitrary, even if the emergence of certain forms of mathematics is a product of historical circumstance. Rather, the mathematics we know on Earth is the creation of human minds as a solution to problems that result from the peculiar ways our mind is embodied. For example, that the decimal system is now the most widely adopted numeral notation on Earth is likely related to the fact that humans have ten digits. An extraterrestrial with ten fingers per hand might find it more natural to operate using the vigesimal system developed by the Aztecs. The point, however, is that whether it is a binary, decimal, or sexagesimal numeral system, these are all metaphorical conceptions of numbers. The important question, as far as the construction of interstellar messages is concerned, is whether we can expect extraterrestrials to have an embodied existence similar to our own and whether we can expect them to be capable of metaphorical thought. I think there is good reason to answer in the affirmative to both queries, but before we consider them in detail, it will be worthwhile to consider the basis of the most prevalent mathematical metaphor in use today, which haunts the mathematical program of every message we've sent into the cosmos: sets.

SET(I) THEORY

In the mid-seventeenth century, the German polymath Gottfried Leibniz set to work outlining the design of a symbolic system that would eradicate all the ambiguities that plagued natural languages. Leibniz hoped his language would be capable of articulating the full spectrum of human thought using only the necessary truths of logic, a dream familiar to the modern architects of interstellar messages. Leibniz was not the first person to undertake this sort of project and although his *"characteristica universalis"* was unsuccessful, it marked a profound improvement over similar projects undertaken by his contemporaries. John Wilkins and George Dalgarno, for instance, sought to create perfect languages based on a collection of primitive elements. Instead of amassing arbitrary primitive elements as the basis of his language, Leibniz realized that the identities of the elements in his calculus didn't matter so much as the relationships between them. It was an insight that would guide George Boole's *Investigation of the Laws of Thought* nearly two centuries later, which marked a turning point in the design of formal logical systems.

Boole's symbolic calculus of human thought effectively conceptualized the notion of classes as numbers and operations upon these classes as arithmetical operators (e.g., union as addition, intersection as multiplication, and so on). From the perspective of embodied mathematics espoused by Lakoff and Núñez, Boole's system capitalizes on a grounding metaphor in which classes are conceptualized as containers. In other words, we intuitively think of the notion of a class as a grouping of objects inside a bounded region of space: those objects that are inside that region are members of the class and those that are outside of it are nonmembers. Boole's logical system linked the metaphor of classes as containers to the metaphor of numbers as objects, a mapping across mathematical domains (classes and arithmetic) that was supposed to reveal the algebraic laws of thought. Boole also added his own inventions to his system— empty classes and universal classes—which are not found in our intuitive, "everyday" concept of classes as containers, but which are critical to the functioning of Boolean algebra. Ironically, the inventions that allowed Boole's system to work were the same that precluded it from being a true calculus of natural human thought.

Boole's metaphorical system, in which classes are conceived as numbers, laid the foundation for symbolic logic and set theory, which would later be claimed as the foundation for mathematics as such (Džamonja 2017). Given that mathematical Platonists were inclined to treat mathematics as an entity woven into the very structure of the universe, it followed that the universe itself could be described in terms of classes, or rather the more technical term, sets. Although symbolic logic and set theory have led to great advances in mathematics on Earth, there's no reason to suppose that this metaphorical system is universal. The empty class and universal class are not a natural feature of human cognition, but a creative solution described by Boole to allow his calculus of thought to work. In short, there is "no scientifically valid reason to believe that the physical entities in the universe form a subclass of an objectively existing universal class" or that the empty class is a feature of the universe and is a subclass of every class (Lakoff and Núñez 2000).

Set theory as developed by Cantor and later axiomatized by Zermelo and Fraenkel uses a more sophisticated version of Boolean classes (Johnson 1972). In Boolean algebra, classes can be subclasses of other classes, but not members of those classes. In axiomatic set theory, classes can be members of other classes. This is an important distinction from the perspective of embodied mathematics. Boolean classes can still adhere to the intuitive "classes as containers" metaphor in which members of a class are conceived as being objects "in" a container, but in axiomatic set theory, the container can contain itself and be treated as an object. Furthermore, set theory treats empty classes as unique objects, rather than the subclass of another class. This is quite removed from our everyday understanding of classes that is grounded in our experience of the world as embodied minds. The consequence of these design changes is that relations and functions can be defined in terms of sets. To take a simple example, consider how numerals can now be defined in terms of sets that have other sets as their members: the empty set Ø can be mapped to 0, the set containing the empty set {Ø} can be mapped to 1, the set containing the set containing the empty set {Ø, {Ø}} maps to 2, the set containing the set containing the set containing the empty set {Ø, {Ø, {Ø}}} maps to 3, and so on.

Axiomatic set theory and symbolic logic have come to dominate mathematics over the past century to the point that it feels entirely natural to conceive of the order of the world in terms of sets and cognition in terms of symbolic logic (Lakoff and Núñez 2000). Yet this misses the critical transition from grounding metaphors, which can reasonably be assumed to be shared with extraterrestrial intelligences since they arise from direct experiences in the world, to linking metaphors, which are the creative products of human minds meant to deal with human experience. In other words, a very useful mathematical invention that meets the idiosyncratic needs of embodied human intelligence has been naturalized to the point that it is taken to be a trait of the universe itself. The consequence of this is that we take it for granted that extraterrestrials will share this conceptual schema because they inhabit the same universe. This is not to say that it is impossible or even unlikely that extraterrestrials will have created functionally equivalent symbolic logics or set theories. Rather, it is a reminder that the ontological status of mathematics is still an area of active debate on Earth and until it is possible to conclusively determine just what mathematics *is*, we must beware of hidden anthropocentric assumptions in even the most ostensibly universal fields like mathematics and the physical sciences.

EMBODIED EXTRATERRESTRIAL INTELLIGENCE

The theory of embodied mathematics based on innate capacity for metaphor raises serious issues about the ontological status of mathematics that are of direct relevance to interstellar communication. If mathematics is a part of the universe that is independent of mind, then we can be relatively certain that extraterrestrials will understand our mathematics. If they are an older civilization than our own, they may have read further in the great book of the universe, but we can rest easy in knowing that we are at least reading the same text. Yet if we abandon mathematical Platonism, we immediately find ourselves in more uncertain territory. If our mathematics is a product of the embodied human mind, then it is perhaps more accurate to say that we are actively writing one version of the great book of the universe from a uniquely human perspective. Although an extraterrestrial is observing the same universe, their interpretation may be much different from our own if their experience as an embodied mind is sufficiently different. Advances in the cognitive and neurological sciences have revealed how the nature of our physical interface with the world—our body—affects our cognition. Thus, it is worth considering whether we can expect an extraterrestrial intelligence to share many physical characteristics with ourselves, which will help inform whether we can expect them to share a similar mathematics.

In some ways, it would almost be more disturbing to make contact with an intelligent extraterrestrial civilization populated by fleshy, mostly hairless hominids than a civilization of eight-eyed cephalopods, but this possibility is not entirely out of the question. Indeed, as the astrobiologist Charles Cockell has argued, empirical evidence suggests that certain features of life are deterministically driven by physical laws. Extrapolating from this, it is reasonable to believe that "at all levels of its structural hierarchy, alien life is likely to look strangely similar to the life we know on Earth" (Cockell 2018). Cockell's argument is analogous to the case made by Minsky that extraterrestrials are likely to think like us because they are subject to the same basic physical constraints. It would be naïve, of course, to suggest that evolution is totally determined by the laws of physics given the significant and obvious role that chance plays in the trajectory of evolution. For example, research suggests that the probability of an asteroid impact resulting in global cooling, mass extinction, and the subsequent appearance of mammals was "quite low" 66 million years ago. It was sheer cosmic bad luck that the asteroid impacted the relatively small portion of the Earth's surface that was rich in the hydrocarbons and sulfur that ultimately choked the Earth with stratospheric soot and sulfate aerosols. In this case, the site of the asteroid impact changed the history of life on Earth in a way that could never be predicted by deterministic evolutionary laws (Kaiho and Oshima 2017).

The point is that although the trajectory of evolution isn't predictable in advance, the variety of species it produces is not boundless. This contradicts the intuitive interpretation of Darwinian evolution, which suggests that natural selection results in a "tendency of species to form varieties" in infinite number. On the contrary, Cockell (2018) argues that "evolution is just a tremendous and exciting interplay of physical principles encoded in genetic material" and "the limited number of these principles ... means that the finale of this process is also restrained and universal." Consider, for example, the emergence of cellular life on Earth. Is the cellular form something that we might expect to emerge on an extraterrestrial planet, or would extraterrestrial organisms find a different mode of self-assembly? In the 1980s, the biologist David Dreamer used carboxylic acids extracted from the famous Murchison meteorite to demonstrate that these simple molecules would spontaneously form cellular membranes when added to water. According to Cockell, this suggests that the ingredients for cellular life are "strewn throughout the Solar System in carbon-rich rocks," which means "we might expect the molecules of cellularity to form in any primordial cloud, ready to deliver their cargo of protocell material to the surface of any planet with a waiting abundance of water." Later experiments demonstrated that meteorites are far from the only source of molecular material that can form cellular membranes, suggesting that this mode of organization is likely common in the universe.

Similar physical laws also limit the possibilities of still more fundamental aspects of biology, such as the structure of DNA. One of the most remarkable features about DNA is that it is composed of only four nucleotides—adenine, thymine, cytosine, and guanine—that can only combine in very limited ways: adenine pairs with thymine and cytosine pairs with guanine. Is the fact that there are only four nucleotides or that they combine into two base pairs an evolutionary accident? Might an extraterrestrial intelligence have a genetic code built from six or more nucleotides, and might these nucleotides be different from the four that comprise the DNA of life on Earth? This is a possibility, of course, but there are strong reasons to believe that it is unlikely.

Adding more nucleotides to the equation increases the amount of information available to the system and means that smaller molecules can contain the same amount of information as longer molecules in genetic pools with only four nucleotides. The trade-off, of course, is that the percentage of bases that a given nucleotide can link with halves with each base pair added to the system. For example, in a two-nucleotide system, each base can pair with half of the bases. In a four-nucleotide system, each base can only link with a quarter of the bases, and so on. Thus, Cockell argues, "as you add more bases, it gets more difficult to find ones that are sufficiently dissimilar to make it easy for them to be distinguished when the molecule replicates," which results in a higher rate of errors. Indeed, computer models of RNA, the molecular interface between DNA and basic proteins, suggest that four nucleotides result in the greatest fitness. As for the types of base pairs, research using synthetic nucleotides to expand the number of base pairs in the genetic code has demonstrated that swapping these synthetic base pairs out of the normal code or adding them usually produces unstable results (Zhang et al. 2017). However, organisms such as bacteria that have synthetic nucleotides added into an expanded genetic alphabet have been shown to be stable under stringent laboratory conditions (Malyshev et al. 2014). The results of ongoing experiments with the many possible base pairs suggest that the four base pairs we see in RNA and DNA are optimized to meet the conditions that allow for its replication, but also the preservation of its structure.

If the brain and its cognitive structures are in fact optimized for the embodied experience of the organism, this suggests that Minsky's thesis that extraterrestrials will think similarly to us is not so far-fetched after all. Extrapolating from this, it is rational to believe that extraterrestrials might hit upon similar mathematical ideas, such as set theory, if their embodied existence is sufficiently similar to our own.

6

TOWARD A *LINGUA COSMICA*

If we are to design a message for interstellar communication, we must consider the strong link between form and content. For example, it is perfectly reasonable to send a television program as a message if we assume that the recipient has an ocular apparatus that allows them to see in the narrow spectrum of visible light. Yet sending a television program on its own is not sufficient if we are to have any hope of the extraterrestrial recipients understanding the content of the message. The way a television image is composed on a screen as an array of dots is an arbitrary matter of convention—the image could just as well be composed from top to bottom or as a spiral out of the center of the screen. Thus, to ensure that the recipient is able to properly compose the image, set an appropriate playback rate, understand the ratio of images on the screen, and so on there must also be a "meta-message" of sorts whose function is primarily didactic—that is, its purpose is to teach the recipient how to decipher the main content of the message.

Obviously, we cannot use a television program to explain how to compose a television image; the meta-message must be composed in a different language, so to speak. Designing this metalanguage and the program for its presentation may be said to be the aim of "astrolinguistics," a term coined by Alexander Ollongren to describe this emerging subdiscipline within the field of METI. To date, there have been only two robust attempts at designing such a language: the first between 1957 and 1960, Hans Freudenthal's Lincos; and over half a century later, Ollongren's revision, based on constructive logic, rather than arithmetic.

Freudenthal was intensely interested in mathematical pedagogy and was resolute that a mathematical education must be rooted in the everyday experience of the pupil. As such it is hardly surprising that Freudenthal became such a fierce opponent of the so-called "new math" that was being taught in the United States and across Europe beginning in the early 1960s. The new math emphasized teaching students to understand the theory behind mathematics without much context or explanation for why such theory was important, which its proponents hoped would endow students with a deeper understanding of mathematical principles. This approach, however, was widely lampooned for producing students who had "heard of the commutative law, but did not know the multiplication table" (Simmons 2003).

Interstellar communication is an extreme instance of mathematics education, and there was hardly anyone more capable than Freudenthal to take on such a task. Although Freudenthal wasn't aware of it at the time, the same year he began to work on Lincos, Lancelot Hogben gave his presentation at the British Interplanetary Society on Astraglossa, his own attempt to design an interstellar message. Despite their independent development, Astraglossa and Lincos share several characteristics, although Freudenthal's program was developed in far greater detail. It is also worth noting that the same year Freudenthal tasked himself with figuring out how to talk to aliens, Chomsky published *Syntactic Structures*, the opening salvo in the cognitive revolution in linguistics in which Chomsky described his formal theory of syntax known as transformational generative grammar. Chomsky's landmark study was published by a Dutch press, but it's uncertain whether Freudenthal was familiar with its contents. If he was, it appears to have had little impact on the development of Lincos. Whereas Chomsky's linguistics emphasized the importance of a formalized syntax in language production, Freudenthal did not attempt to develop a full syntax for Lincos and eschewed formalization wherever possible. Indeed, Freudenthal (1960) wrote that his "purpose is to design a language that can be understood by a person not acquainted with any of our natural languages or even their syntactic structures." Freudenthal believed that it is possible to overcome the challenge of communicating with extraterrestrials who have no knowledge of our language because it is comparable to "a problem that is daily solved in the intercourse with babies and infants. We succeed in teaching them our language though we have started with a tabula rasa of lexicologic and syntactic knowledge." While Freudenthal is certainly correct insofar as a lexicon must be conveyed to a child through examples, as we saw in earlier chapters infants are far from a "tabula rasa" when it comes to syntactic knowledge; rather, they are born with the hardware necessary for language. As Chomsky theorized and recent work in neurolinguistics continues to confirm, there is a fundamental connection between the structure of our brains and our ability to produce language, so it's quite unlikely that we would be able to converse with an extraterrestrial whose language is not structured by the same universal grammar. While this doesn't invalidate the Lincos project since Freudenthal still assumes that "the person who is to receive my messages is human or at least humanlike as to his mental state and experiences," it may have limited its full development insofar as it lacks a complete syntax.

Freudenthal's Lincos is to be considered a type of speech rather than writing; Lincos words are made up not of letters but of phonemes, which consist of unmodulated radio waves varying in wavelength and duration. Freudenthal did not bother to explain how these phonemes would be "pronounced" when broadcast as radio waves, although this problem was addressed for the 1999 Cosmic Call message, which was based on the Lincos program. When Freudenthal published *Lincos* in 1960, he did not intend for it to be a full program that was ready to be broadcast to the stars. Rather, it was more of a general outline for such a program, which would include far more examples than Freudenthal included in his monograph. The text was meant to be the first entry in a two-part series, although the second part was never published. The first book was divided into chapters dealing with mathematics, time, behavior, space, motion, and mass, whereas the second volume would cover matter, Earth, and life. As seen in the progression of these topics, Lincos begins with very abstract concepts before moving to more concrete information about life on Earth, which is conveyed through "morality plays." These parts of the message depict humans interacting with one another to teach ideas about human behavior and are conveyed by borrowing heavily from mathematical syntax.

At first blush, the idea of beginning an interstellar message with basic arithmetic seems overly pedantic. After all, if extraterrestrial recipients can build a radio receiver there is little reason to doubt that they are capable of doing basic arithmetic. So why did Freudenthal start with teaching basic math? The main reason

is that numbers are ostensive—they are, in Freudenthal's termi-
nology, "ideophonetic," in the same way a hieroglyph depicting an eye which also refers to an eye is "ideogrammatic." Aside
from units of time, numbers are the only concepts in Lincos that
can be directly understood insofar as one pulse means "one,"
two pulses mean "two," and so on, and the notion of one second
can be demonstrated by a pulse lasting one second. By beginning with signs that mean nothing but themselves, Lincos can
establish a common ground with the extraterrestrial recipient in
order to introduce words whose meanings must be determined
based on their relationship to other words in an utterance. For
example, once "one," "two," and "three" have been introduced
through ostensive pulses, it is possible to introduce words for
addition, subtraction, greater than, less than, and equal to with
elementary formulas. In other words, the system is not teaching
the recipients math so much as it is teaching them the Lincos
lexicon. Once sufficent common ground is established on the
basis of knowledge that can be reasonably assumed to be common to Earthlings and extraterrestrials (e.g., elementary mathematics and physics), it is possible to convey contingent facts
about human customs.

From the very first examples in Lincos, we can see Freudenthal's emphasis on the communicative aspect of Lincos. Rather
than provide formal axiomatic definitions for mathematical
ideas (e.g., the commutative, associative, or identity properties
of arithmetic), Freudenthal instead chose to convey these ideas
through "quasi-general definitions," where an idea is conveyed
through a sufficiently large number of examples so that an intelligent receiver can generalize from the examples to arrive at the
definition. He argued that this is preferable because it is more
in line with how math is learned by children, who generalize

from a relatively small number of examples and some informal rules. This was contrary to the dominant trends in mathematics in the early twentieth century—best exemplified in the work of Alfred North Whitehead and Bertrand Russell—which sought to ground the foundations of mathematics in formal axiomatic systems (Russell and Whitehead 1925). In these formal systems propositions were written using logical terms. The aim of reformulating propositions in a logical language was to allow the truth-value of statements to be determined regardless of the meaning of the phrases themselves. In this sense, Russell's logical language is more like a code than a natural language, insofar as its elements can be manipulated according to a set of rules without regard for the meaning of the expressions themselves. It is perhaps unsurprising, then, that the work on logical language found its most significant applications in the development of computer languages rather than in clarification in day-to-day human communication.

Although Russell's logical language dealt with logical problems that arise in vernacular language, it was never meant to fulfill a communicative function, which Freudenthal viewed as the defining characteristic of language. In Russell's atomistic approach to language, he considered elementary linguistic "atoms" to be endowed with a definite meaning, and when these atoms are combined into larger phrases the truth-value of the resulting proposition can be determined based on the logical relations of the atoms contained therein. Freudenthal rejected this, writing that he "cannot imagine general rules of meaningfulness," and that he does not "know how to attach just one meaning to every meaningful linguistic expression." Instead, when language is viewed as a communicative tool, rather than an isolated thing to be studied on its own, he argued that the meaning of a proposition is dependent on the context. Although some formal aspects of Russell's logical language were kept in Lincos, Freudenthal rejected a fully formalized language for interstellar communication and dispenses with formalization wherever possible. "Our dissenting from formalist semantics is only a consequence of our dealing with language as a means of communication," Freudenthal wrote. "A language such as Russell's in which 'Walter Scott' and 'the author of *Waverly*' may be freely substituted for each other cannot serve our purposes because it does not allow to communicate the *fact* that Scott is the author of *Waverly* and to ask the question whether Scott is the author of *Waverly*" (Freudenthal 1960).

On the other hand, Freudenthal argued that vernacular languages would also be insufficient grounds for the syntax of an interstellar language, given their inability to handle variables and punctuation in a clear, systematic way. Thus, Freudenthal saw mathematics as occupying something of a middle ground between the rigid formalism of logic and the ambiguity of natural language. Mathematics provides the best of both, insofar as it systematically deals with variables and punctuation, yet can be communicated with quasi-definitions and can reasonably be supposed to be universal.

COSMIC CALLS

Freudenthal never lived to see his *lingua cosmica* used in an interstellar broadcast, although Lincos was apparently used on Earth. At the 1971 CETI meeting in Armenia, Carl Sagan mentioned that some testing of the language's effectiveness had been done with school children in the Soviet Union, but that the results of these tests indicated that Lincos was perhaps not as intuitive as Freudenthal had hoped. Nevertheless, shortly after Freuden-thal's passing, his contribution to exolinguistics was used as the basis for the first scientific interstellar broadcast from Earth since the Arecibo transmission in 1974. In 1999, the Cosmic Call mes-sage was broadcast from Ukraine's Evpatoria Deep Space Center, home to the second most powerful planetary radar on Earth at the time, to four stars between 51.8 and 70.5 light years away. The targeted stars were selected based on their proximity to the galactic plane, which meant that the message could reach an additional ten Sun-like stars that weren't considered primary tar-gets, as well as their similarity to our own Sun in terms of metal-licity, age, and spectral type.

The Cosmic Call broadcast was overseen by a short-lived company called Team Encounter created by the serial space entrepreneur Charles Chafer. In 1998, Chafer gave the go-ahead to two Canadian physicists, Yvan Dutil and Stephane Dumas, to spearhead the design of the Cosmic Call message after they reached out to the company with ideas about how to construct a scientific message. As self-described SETI hobbyists, Dutil and Dumas had participated in an internet project to decipher hypo-thetical extraterrestrial transmissions created by Team Encoun-ter, yet the design of their Cosmic Call message was informed by Freudenthal's *Lincos*. Freudenthal hadn't gone so far as to actu-ally create a program for transmission, however, so it was up to Dutil and Dumas to translate Freudenthal's insights into a work-able message.

Freudenthal envisioned Lincos as a spoken language whose phonemes would be encoded in unmodulated radio waves. This design choice was rejected by Dutil and Dumas in favor of a bit-map on the grounds that it would be more robust against noise and decoding errors. Although bitmaps offer greater redundancy, they are also susceptible to corruption in interstellar space. Changing a single bit in the 1974 Arecibo message, for instance, could destroy the entire structure of the message. To avoid this, Dutil and Dumas create a one-bit "frame" around each of the 23 pages in their message to create a larger structure that would provide a point of reference for an extraterrestrial reconstructing the message. Each page of the message was a 127×127 bit array (127 is a prime number although the product of bitmap dimensions, 16,129, is not) that was populated with 5×7 bit ideographic symbols created by Dutil and Dumas that are all unique even when rotated or mirrored (see appendix B for a complete list of the symbols). Like Freudenthal's message, the Dutil–Dumas component of the Cosmic Call message began by introducing the numerals 0 through 9 and basic principles of arithmetic through a handful of examples. This established the basis for the introduction of increasingly complex topics. Included in the message were "lessons" in geometry, units of measurement, the chemical elements, the basic characteristics of our solar system and Earth, a facsimile of the male and female human depicted on the Pioneer plaques, the structure of DNA, a Fuller map of the Earth, the characteristics of the Evpatoria telescope, some basic cosmological concepts such as the age of the universe and its temperature, and finally some questions asking the recipient for information about its own planet and civilization.

In addition to the Dutil–Dumas message, the Cosmic Call transmissions also included a shorter scientific message designed by the Team Encounter employee Richard Braastad, which used the symbol system developed by Dutil and Dumas to describe a light sail spacecraft that was under development by Team Encounter at the time. Also included in the transmission were the 1974 Arecibo message, personal messages from the Team

Encounter staff, and a public message. The four scientific messages were each broadcast three times in a cycle that began with the Dutil–Dumas message, followed by the Braastad message, the Arecibo message, and the Team Encounter staff message to ensure enough redundancy in the event that portions of the message were corrupted in transit. The transmission concluded with the public message, which was broadcast only once and consisted of the names and a short message submitted by over 43,000 people from around the world.

On May 24, 1999, the first Cosmic Call message was broadcast from Evpatoria to a star in the Cygnus constellation approximately 70 light years from Earth. The message was transmitted at 100 bits per second (except for the public portion, which was transmitted at 2,000 bits per second) on the 5.01 GHz (6 cm) band. (This isn't a "magic" frequency but was the frequency at which the radio telescope was equipped to broadcast.) The transmission used frequency-shift keying to modulate the radio signal, which shifted the transmission frequency by 24 kHz to encode the message as a ternary stream. The message itself was encoded in binary, such that 0 was represented at 5,010,000 kHz, 1 represented at 5,010,048 kHz, and a five-second pause between each of the messages was represented at 5,010,024 kHz. Team Encounter had planned to repeat the broadcast to three other stellar targets the following evening, but these plans were thwarted by the National Space Agency of Ukraine, which had been spooked by the public response to the message. After learning that the United States' radio observatories had refused to transmit the message and receiving an overwhelming number of press inquiries wondering if the transmission was dangerous, the agency pulled the plug on the project. It was only after Alexander Zaitsev, the Russian radio astronomer spearheading the transmission, made a trip to Kiev to convince the agency that the project wasn't dangerous that transmissions were allowed to resume to the three other stars the following month.

It's difficult to overstate the importance of the 1999 Cosmic Call message in reigniting interest in interstellar communication. Team Encounter demonstrated that scientific broadcasts could be financed through small donations and that there was significant public interest in the possibility of communicating with extraterrestrial life. The decade following Cosmic Call saw a surge in the number of interstellar broadcasts, many of which were blatant money grabs or corporate publicity stunts, but the pursuit of scientific interstellar messages was far from dead. On July 6, 2003, Team Encounter sent its second Cosmic Call message from the radio telescope at Evpatoria to five stars that had similar characteristics to our own Sun and were between 32.8 and 45.9 light years from Earth. The content of the second Cosmic Call message was similar to the first, but a few key changes involving the format and transmission of the message were made to increase the chance that it would be correctly interpreted. The first major change was merging the 23 pages of the original message into one long page, which was a more efficient use of space and also made the message more likely to be successfully decoded since vertical lines are more important than horizontal lines in the decoding process. Moreover, the 5×7 symbols were reduced to 4×7 symbols, which Dutil and Dumas found to be more resistant to noise. The 2003 Cosmic Call message also included some novel components that were absent from the 1999 message, such as a Bilingual Image Glossary, the Ella chatbot, and several images. The Bilingual Image Glossary consisted of twelve 101×101 bitmaps that depicted drawings of things like children, Earth, a game of tic-tac-toe, and the Sun, with the name of each picture written in English and Russian. (This Bilingual Image Glossary was developed as part of the Teen Age Message sent from the Evpatoria telescope in 2001, which will be discussed in greater detail below.) The transmission speed of the scientific portions of the 2003 Cosmic Call message was also increased from 100 bits per second to 400 bits per second. The public portion of the message consisted of 90,000 individual messages transmitted at 100,000 bits per second, which is over fifty times faster than in 1999. Nevertheless, the personal messages alone took about eleven hours to transmit. Rather than transmitting each individual portion before repeating the entire call as was done in 1999, the 2003 Cosmic Call message repeated the Dutil–Dumas message, the Arecibo message, and the Bilingual Image Glossary three times each before transmitting the Braastad message, the Team Encounter staff message, and the public messages one time per star.

The Cosmic Call messages won't start arriving at their target stars until 2036; if there's anyone listening there that is able to decode the message, then the earliest we can expect a reply is around 2070. Although these messages represent significant progress in the art and science of interstellar message construction, they left plenty of room for improvement.

LINCOS 2.0

Freudenthal was explicit about his rejection of formalism wherever possible on the grounds that it did not fulfill a communicative function. As such, his Lincos oscillates between the formalism of logic and the communicative style more akin to natural languages. Rather than formally defining a number using something like Peano's postulates, for instance, Freudenthal opted to introduce numbers by way of example. This makes Lincos something of a chimera in the context of historical attempts at designing a perfect language, which almost always fell on the side of extreme formalism (Bassi 1992). Yet the unique purpose of Lincos also meant that it was burdened with restrictions that were not applicable to other perfect languages. Whereas languages meant to clarify the exchange of ideas between Earthlings could be explained in a natural language shared by the teacher and pupil and the meaning of terms could be clarified through feedback, none of these advantages are available in an interstellar context. In cosmic communication, the extreme distance between sender and receiver, coupled with an upper bound on transmission speed, makes real-time feedback impossible, and the applicability of ostensive definitions are limited to concepts like numbers, time, and if we have knowledge of the stellar target's planetary system, perhaps some celestial events.

These stringent restrictions meant that Lincos must be taught in terms of itself. Freudenthal hoped to draw on the best aspects of formal languages and natural languages to create a hybrid system that would be optimized for communication with a species that is perhaps radically different from our own in many respects, but still shares some crucial features with humans, such as having a language and similar intuitive concepts about arithmetic. From formal languages, Freudenthal borrowed the principle of using punctuation to organize syntax and the treatment of variables, which can range over any domain rather than a unique domain determined by the lexicon. Despite this moderate formalization, other aspects of Lincos are far less systematic, and this is often to the system's detriment. For example, there is no systematic treatment of semantic and syntactic categories. Instead, new words are introduced ad hoc and their role in the syntax is not defined formally but must be grasped by way of example. Moreover, question marks are used to bind variables, but do not elucidate the *meaning* of a phrase. The result, then, is that questions employing words like "why" or "whether" (which don't bind variables in a requested answer) can't be differentiated from affirmative statements based on the sentence's syntax. These shortcomings do not appear to be an oversight by Freudenthal so much as an intentional design choice, a way to prioritize the communicative aspects of natural languages in his *lingua cosmica*. The result, however, is a language that relies on a great deal of creative interpretation from the recipient, who will still be left without any way of verifying that their interpretation is the correct one. If Freudenthal is correct, a large degree of ambiguity may be an unavoidable feature of any *lingua cosmica* that can be used to communicate meaningfully, as opposed to just manipulating meaningless symbols to communicate logical truths.

What Freudenthal couldn't foresee, however, was that advances in computer science in the second half of the twentieth century made it possible to use a strongly formalized language to communicate meaningfully and in such a way that the correct interpretation of the message could be guaranteed. Indeed, insights from computer science are at the heart of the second generation of Lincos developed by Alexander Ollongren and serve as the basis for a new research paradigm he called "astrolinguistics." Similar to how Chomskyan linguistics sought universal grammatical features common to all terrestrial languages, Ollongren's astrolinguistics aims to discover whether there are any general linguistic rules that are applicable to all intelligent species in the universe, as well as the degree to which our conceptualization of language influences our understanding of astrophysical phenomena. Ollongren (2013) describes language in broad strokes as "a basic set of some kind of tokens together with rules for the formation of possible expressions (the *syntax* of language)—but also additional rules for the use and goal of these expressions (the *semantics* of language supply *interpretation*)." Looking to astrophysical processes in the cosmos, Ollongren saw similar principles at work in the interactions of elementary particles, such that "packets of energy in various forms can be considered to be the basic tokens of a linguistic principle on a cosmic scale" and "the syntactic rules are those governing interactions between the tokens." Yet Ollongren's astrolinguistics also aims to explore "whether linguistics of natural languages on Earth is some sort of derivative of a general cosmic principle, valid for all living intelligent beings in our universe." Until we make first contact, the existence of any truly *universal* linguistic features is pure conjecture. In any case, Ollongren raises an important question: which features of human language can we reasonably expect to be present in an extraterrestrial language? Ollongren acknowledged that it is possible that human languages have developed according to the limits of the human brain, but he nevertheless sees the ability of human languages to explain the rules governing well-formed expressions within the language itself as a critical feature of human languages that might reasonably be considered as a general linguistic principle throughout the cosmos.

Building upon this insight, Ollongren designed a *lingua cosmica* that is, first and foremost, capable of self-interpretation. Ideally, this language would consist of the "simplest possible grammatical structures," yet would also have the expressive power to describe any of the information we desire to transmit. Ollongren is a generation younger than Freudenthal and as such came of age at a time when the first procedural and functional programming languages such as ALGOL and LISP were being developed. According to Ollongren, his familiarity with these programming languages gave him the idea that "Freudenthal's abundant use of super- and subscripts and the numerous *ad hoc* agreements might be simplified with some effort by using ingredients from 'modern' theories of computer programming." He also realized that "the overall purpose of Freudenthal's work might be achieved in a better way if one *abstained* from using just a single level in interstellar message construction—that of the *lingua cosmica* itself" (Ollongren 2013) And so, in the late 1990s, Ollongren set to work designing a second-generation Lincos that was multileveled and based on logic rather than mathematics. For Ollongren, logic has a distinct advantage over mathematics as a basis for a *lingua cosmica* because it can describe the logical content of a text, as well as the definitional framework of the system. This is similar to the way an environment is created in computer programs through declarations, rather than relying on the intuition of the recipient to reconstruct this framework based from provided examples. There are many different kinds or modalities of logic, all of which are designed to lead to "correct reasoning over abstractions of reality," but Ollongren notes that they are not all equally useful in their ability to clarify an argument. For example, propositional logic has a relatively simple syntax in which propositions are combined using a small set of logical operators (i.e., and, or, negation, and implication) and the truth-value of a statement can be simply determined through a truth-table. Propositional calculus is only able to abstract over individual objects, however, which means it cannot be used in situations requiring abstractions over sets of objects satisfying some predicate. Predicate logic overcomes this problem by introducing the quantifiers "all" and "exists," but this comes at the cost of a more complicated syntax. Consider a case where all the objects in some set have some feature in common. In predicate logic, it would be false to say that an object possessing this feature exists "unless such an entity existed already or is constructible with instruments of the logic." The calculus of constructions with inductions (CCI) addresses this dilemma by providing expressions with types, "either by introductory declarations or by simple construction and reduction rules." This overcomes the existence problem described above by declaring a constant entity that has the type shared by the objects in the set. This logical paradigm is adapted from the lambda calculus, which has the advantage of an extremely simple syntax and a very small set of primitives. Still, CCI is quite expressive and can convey logical descriptions of dynamic situations, such as human behavior (see appendix D for an introduction to the lambda calculus and its application to Lincos).

Like Freudenthal, Ollongren made no attempt to design a program for interstellar transmission. Instead he attempted to show how the logical content of a text could be conveyed with constructive logic. His second-generation Lincos is a system consisting of at least two levels. The first level consists of a message, which may consist of text, images, or music. The second level consists of a metalanguage that is a formalized annotation system to describe the logical contents of the primary text. This is very similar to the process of textual annotation for natural language processing applications in computational linguistics, in which metadata is grafted onto a natural language text to give a machine learning algorithm extra information that allows it to "understand" the way language functions in the base text.

Ollongren's Lincos draws on insights from machine learning techniques insofar as the language annotates a natural language text (in the broad sense—this may also include images or music). It is also functionally similar to Coq, a proof-checker developed in the late 1990s based on the CCI, which guarantees that expressions in Lincos will always be correct and makes the language self-interpreting. Still, the system is burdened with issues of how expressions will be interpreted by the recipient. For example, it is unclear how to create a signature to convey the modality and the "semantics" of the logical system being used. As for issues of semantics, Ollongren notes that Lincos is essentially a way of interpreting other information (i.e., the base text), but what if Lincos as such were interpreted by an extraterrestrial automaton? It's uncertain how the automaton would interpret the message, since Lincos doesn't describe operations or actions the way a computer program does, but instead describes logical relations between terms. At best, the automaton would arrive at a relational analysis of essentially meaningless terms. In other words, Lincos wasn't designed for interpretation by an automaton, but rather for an extraterrestrial intelligence with "knowledge of the world." Exactly what knowledge should be transmitted, however, is a difficult decision that Ollongren leaves for the designers of interstellar messages.

7

HOW TO TALK IN SPACE

An issue that is intimately linked with the design of a *lingua cosmica* and the contents of an interstellar message is that of the transmission medium. Throughout this book we have tacitly assumed that the mode of transmission is electromagnetic radiation in the microwave range of the spectrum, but this is certainly not the only option. On Earth, different modes of communication each come with their own trade-offs in terms of energy requirements, risk, and the types of information they carry. For example, using the postal service is incredibly resource intensive compared to sending an email, but only the former mode of communication has the capacity to transport a physical object between sender and receiver. The same is true of cosmic communication modes, which address the unique constraints of communicating in space with varying degrees of success. Ultimately, the choice of communication mode depends on what the sender wishes to communicate and the availability of resources.

PHYSICAL MEDIA

To date, only four physical objects are destined for interstellar space that can be considered candidates for extraterrestrial communication. The first to launch were plaques affixed to Pioneer 10 and Pioneer 11. These probes were launched by NASA in 1972 and 1973 to study Jupiter, solar winds, asteroids, and cosmic rays. In addition to being the first spacecraft to visit the outer reaches of our solar system, the Pioneer craft had the notable distinction of being the first to attain a velocity that would allow them to leave the solar system, a fact whose significance was not lost on the science journalist Eric Burgess. Just months before Pioneer 10 was scheduled to launch, Burgess approached Carl Sagan and suggested that NASA include a message on the craft in case any extraterrestrials happened upon it while it was traversing interstellar space. NASA assented and gave Sagan three weeks to design the message. The resulting message, which was designed with input from Frank Drake, consisted mostly of some rudimentary scientific information that they hoped would help extraterrestrials locate and communicate with Earth if the craft was intercepted. The contents of the message were provided to Sagan's wife, the artist Linda Salzman Sagan, who furnished the drawings and diagrams that adorned a small 6×9-inch plaque attached to Pioneer 10.

The key to the Pioneer message is a schematic depicting the spin-flip transition of a hydrogen atom in the upper-left corner of the plaque. A hydrogen atom consists of a proton and electron, each of which has a magnetic dipole moment—envision each particle as a magnet whose orientation can be either "up" or "down"—as a result of the particle's spin. If the electron's magnetic dipole moment is parallel to the dipole moment of the proton it increases the energy in the atomic system and decreases the energy if they are antiparallel. Approximately every 0.7 nanoseconds, the electron undergoes a spin-flip transition that reverses its magnetic orientation. This transition between energy levels produces microwave radiation at a frequency of approximately 1,420 MHz, which is equivalent to a wavelength of about 21 centimeters. On the plaque, the binary digit "one" stands between the two states of hydrogen, but based on this schematic alone it is unclear whether this is meant to convey a unit of length (21 cm) or a unit of time (1,420 MHz). To resolve this issue, Sagan and Drake included the binary representation of "eight" between two lines and next to a scale drawing of the Pioneer spacecraft in profile and a male and female human. This is meant to act as a sort of ruler for the extraterrestrial, who would be able to measure the spacecraft and find that 168 centimeters (eight multiplied by twenty-one) corresponds to the height of the spacecraft depicted on the plaque. This would have the added benefit of conveying the average size of humans on Earth, since they are depicted to scale alongside the Pioneer craft.

Sagan and Drake's choice of the hydrogen atom was deliberate. As the most abundant element in the universe, the spectral line of neutral hydrogen was considered at the time to be the most promising frequency to search for an extraterrestrial message. By calling attention to this radio frequency, Sagan and Drake were effectively giving an extraterrestrial recipient Earth's phone number. If they were to call on that frequency, someone would probably be listening. Yet a phone number doesn't do much good if you don't know the area code, and the same goes for interstellar communication. To this end, the Pioneer plaques include two maps to assist an extraterrestrial in locating Earth. One of the maps depicts the nine planets of our solar system (Pluto was still dignified with planetary status in 1972) and the Sun. An arrow drawn from the third planet and looping around the fifth depicts the *Pioneer*'s trajectory from Earth around

Jupiter. The relative distance between the planets is labeled in binary. The second map consists of fifteen lines all emanating from a common origin. Fourteen of these lines consist of a long binary number that corresponds to a ten-digit number in decimal notation. These lines correspond to the distance of the Sun from fourteen different pulsars; the fifteenth line corresponds to the distance of the Sun from the center of the galaxy. Yet what is an extraterrestrial to make of the binary numbers? Once again, it appears that these numbers could represent either a time interval or a length.

Sagan and Drake presumed that an extraterrestrial would know that our civilization is unable to measure the distance to the nearest stars to a ten-digit accuracy based on the level of technical sophistication demonstrated by the Pioneer craft themselves. However, it would be plausible that a civilization like ours could measure time intervals to this accuracy. Thus, Sagan and Drake presumed it would be obvious that these time signatures corresponded to the intervals of various pulsars, which function like "galactic clocks." Pulsars are rapidly spinning neutron stars that appear to pulse because they are effectively emitting a beam of radiation that sweeps an area of the sky at regular intervals that hardly vary over the course of millions of years. Sagan and Drake selected fourteen young pulsars with short periods that were evenly distributed in the galaxy so that there would be a good chance that at least some of them would be able to be detected by an extraterrestrial civilization. Given the regularity of the pulsars, they presumed that an extraterrestrial would be able to determine Earth's approximate location relative to these pulsars and how long ago the Pioneer 10 craft was launched.

Although an identical plaque was attached to Pioneer 11, which launched eleven months after Pioneer 10 on a mission to study Saturn, neither Sagan nor Drake harbored any illusions that the Pioneer spacecraft would end up in an alien solar system hosting a technologically advanced extraterrestrial civilization. Both spacecraft are currently headed out of the solar system at around 25,000 miles per hour, but even at these blistering speeds it would take the Pioneer probes around 80,000 years to reach our closest stellar neighbor, Alpha Centauri. Alas, the Pioneer probes are not headed in the direction of the Alpha Centauri system, which led Sagan and Drake to estimate that the mean time it would take for the probe to come within 30 astronomical units of another star is longer than the age of the Milky Way. Despite these poor odds, Sagan and Drake considered it "possible that some civilizations much more advanced than ours have the means of detecting an object such as Pioneer 10 in interstellar space, distinguishing it from other objects of comparable size, but not of artificial origin, and then intercepting and acquiring the spacecraft" (Sagan, Sagan, and Drake 1972).

The content of the Pioneer plaques has been criticized on several grounds, such as the depiction of nudity, sexual bias, and the racism of using human figures that looked Caucasian. These are valid concerns that will need to be addressed when constructing any message for interstellar communication, but these critiques also presume that an extraterrestrial intelligence would be able to understand the contents of the message at all. Sagan and Drake admitted that the message on the Pioneer plaques was "an adequate but hardly ideal solution" and that "any such message will be constrained to a greater or lesser degree, by the limitations of human perceptual and logical processes." Even though the message "inadvertently contains anthropocentric content,"

Sagan and Drake still felt that any advanced technological civilization that might encounter it would be able to decipher its contents. Obviously, the message assumes that its recipient will have some sort of visual faculty, but this is hardly sufficient to interpret the symbols on the plaque.

In 1977, just four years after Pioneer 10 completed its historic flyby of Jupiter, NASA launched Voyager 1 and Voyager 2 probes on a mission to do flybys of Jupiter, Saturn, Neptune and Pluto. During their mission, the Voyager spacecraft attained velocities over 35,000 miles per hour, more than enough to escape the solar system. It was a perfect opportunity to construct another message for any extraterrestrial civilizations that might intercept the craft. Sagan once again spearheaded the project, but this time around he had far more time and resources to craft the message. The result was a pair of gold-plated phonographic records. On the cover of each record are several diagrams, including the same hydrogen transition schematic and pulsar map from the Pioneer plaque. There was also a diagram of the record itself that showed the correct placement of a stylus to play the record and a number expressing the correct rotational speed of the record (3.6 seconds) that is expressed in terms of the time period of the transition of the hydrogen atom (0.7 nanoseconds). The record contained a variety of recorded sounds, including Beethoven, traditional West African Music, Sagan's laughter, bird song, and greetings in 55 different languages, as well as 116 pictures. Any extraterrestrial intelligence that happens upon the *Voyager* is unlikely to know how to extract the visual data contained on the record, so the cover also contains instructions for how to reconstruct the image and a drawing of the first image on the record (a calibration circle) for reference.

The Voyager records were a great improvement over the Pioneer plaques in many respects. This is particularly true in terms of the diversity of their content, which is a closer approximation of human diversity on Earth. The record also contains far more information, although it's far from certain this information would be sensible to an extraterrestrial intelligence if it managed to extract the data from the record. Phonographic records are elegant storage devices that record audio data using a device that converts sound waves into electrical energy and finally mechanical energy, which is used to create grooves in some medium (typically vinyl) whose varying depths correspond to the sound wave. To play the data back, a needle attached to a transducer converts the mechanical energy produced as the needle tracks the groove depth into electrical energy, which is then converted into sound waves. The ability to convert one form of energy into another form of energy is fundamental for any technologically advanced society, and it can be assumed that an extraterrestrial intelligence capable of intercepting the spacecraft has developed transducers. If they have not, we can't hope to communicate with them at all because they won't be capable of radio astronomy. A close analysis of the Voyager records would reveal that the depths of the grooves in the record are variable and when this variation was plotted it would look like a waveform, giving some indication that the grooves are used to encode data.

If the extraterrestrial does manage to extract the audio from the record, however, it will likely be meaningless because it is unlabeled. So, while there are greetings in 55 different languages, there is not enough linguistic data in any given language to perform a pattern analysis of its contents. Moreover, the greetings differ in their contents, so if the extraterrestrial were to treat the audio as a sort of cosmic Rosetta Stone and attempt to find relationships among the greetings, they will be misled from the start. At best, one may hope that the extraterrestrial is merely able to tell that the greetings are linguistic data based on the statistical relationship between the lexical units. As for the non-linguistic sounds, one may assume that the extraterrestrial will be at a loss to decipher their significance. What would an ETI make of the roar of an F-111 screaming overhead? It may be able to determine that this is not a sound made by an animal because its frequency range is too broad, but will it link the aircraft's acoustic signature to a sophisticated weapon? Things are more hopeful for the animal sounds, and particularly the chirping of birds and the moans of whales. Using an "intelligence filter" of the kind envisioned by Laurance Doyle, an extraterrestrial may be able to tell that the chirps convey information but are not the same type of communication system as the languages used in the greetings. Unfortunately, each of these animal sounds lasts for only a few seconds and may not provide enough data to make an analysis of the frequency of meaningful units in these animal communication systems.

The 116 images included on the record begin with six slides developed by Drake to describe some basic mathematical defini-tions, physical concepts, and aspects of our solar system. The first six natural numbers are introduced with an ostensive defi-nition (a row of dots) and linked to their respective binary and Arabic numerals, thereby introducing the arithmetical notion of identity. This provides the basis for introducing other basic arithmetical operators such as addition and multiplication by way of simple equations such as "$2 + 3 = 5$." Next, basic units of measurement for time (second, day, year), mass (gram and kilogram), and distance (centimeters, meters, kilometers) are defined in terms of the period of hydrogen transitions and the mass of a hydrogen atom. The definitions introduced in these first images are used to label other pictures in order to provide context so that an extraterrestrial can learn interesting information, such as the diameter and mass of various planets in our solar system.

Despite their best intentions, the Voyager record and Pioneer plaques are each a sort of cosmic message in a bottle that is unlikely to ever wash up on alien shores. Even if the craft were intercepted, the deep anthropocentrism of each message makes it unlikely that their contents would be correctly deciphered. Indeed, Sagan later acknowledged that these artifacts were designed more for the sake of Earthlings than serious attempts at facilitating interstellar communication. Physical messages are an incredibly energy-efficient mode of communication in terms of the amount of information conveyed per unit of energy required to send a physical artifact to a given destination (Rose and Wright 2004), and, since the launch of the Voyager records, numerous further proposals for message-bearing interstellar probes have been put forward. At this point in time, however, the sheer number of potential targets for a probe makes their deployment to other solar systems an impractical proposition.

MICROWAVE METI

Except for the Pioneer plaques and Voyager records, every interstellar message has used microwave frequencies as the communication medium. Although the microwave band ranges from 300 MHz to 300 GHz, METI mostly focuses on a relatively narrow band ranging from about 1 GHz to 10 GHz. This is known as the "microwave window," so called because it is one of the few regions of the electromagnetic spectrum that is unimpeded by Earth's atmospheric gasses. At frequencies below 30 MHz, the ionosphere is opaque, meaning these frequencies cannot be broadcast from Earth into space or received on Earth from space. The atmosphere is also opaque at most frequencies higher than 300 GHz, apart from a few bands that allow visible light and some ultraviolet rays to pass through the atmosphere. Earth's microwave window technically extends up to 30 GHz, but the 1GHz to 10 GHz band is especially prized by radio astronomers and SETI researchers because it is also the quietest part of the radio spectrum in interstellar space. The universe is awash in radio noise created by astrophysical phenomena such as the cosmic microwave background, quasars, pulsars, nebulae, and even the planets in our own solar system (Horner 1957), but galactic noise starts to become particularly "loud" at frequencies below 1 GHz. Since the strength of radio signals is beholden to the inverse square law—the strength of the signal is inversely proportional to the distance from its source—and the signal will have to travel through energy-absorbing clouds of interstellar dust and gas, any intelligible signal sent to an extraterrestrial civilization from Earth will be quite faint by the time it arrives at its destination. To maximize the chance that an ETI will be able to detect an interstellar message among the cosmic noise it's necessary to broadcast in the quietest band of the radio spectrum and on a frequency that is likely to be monitored by the recipient.

Since most astrophysical noise is smeared across a very wide band of frequencies, the conventional wisdom is that a broadcast on a single, narrow carrier frequency is more likely to stand out as intelligible signal to any extraterrestrials that may be listening (Shostak 1995). Modern SETI programs can scan millions of narrow channels (e.g., around 1 Hz) in real time and up to billions of channels offline. Monitoring narrowband channels reduces the background noise because the channel represents only a small portion of the broadband noise, while at the same time "cranking up the volume" of any signal intentionally broadcast on that frequency. There is good reason to suspect that any extraterrestrial radio astronomers would adopt similar narrowband search strategies. The question for METI, then, is how narrow to make the signal. In the early days of SETI, the assumption was that narrower was always better. This is still true in general, but only up to a quantifiable point. When radio signals pass through clouds of free electrons in interstellar space, the motion of these electron clouds induces a Doppler effect on the signal, which curves the path of some of the photons. The photons whose paths are curved take longer to arrive at the receiver, which has the effect of increasing the bandwidth of even the narrowest signals (Cordes and Lazio 1991; Drake and Helou 1977). This defines a lower bound for frequency bandwidth in interstellar communication, which is around 0.1 Hz and known as the Drake-Helou limit (Horowitz and Sagan 1993). Nevertheless, it is still possible to distinguish signals that have been affected by this type of dispersion as artificial. For instance, a signal broadcast at 1,420 MHz will only see a spread of about 30 MHz Similarly, interstellar scattering sets an upper limit on broadband signals at around 200 kHz (Shostak 1995).

Although the microwave window represents less than 1 percent of the microwave band, it still offers a staggering array of frequencies from which to choose. If an interstellar broadcast has a bandwidth of only 1 Hz, which is approximately ten times greater than the absolute minimum, then the extraterrestrial can still choose from nine billion possible frequencies. If the extraterrestrial's civilization highly values SETI, it's possible they are monitoring all the frequencies on the microwave spectrum. If their civilization is like our own, however, then their SETI programs will be constantly on the verge of running out of funding and their most sophisticated programs will only be capable of monitoring around 1 percent of the spectrum in a narrow portion of the sky at any given time. If we take our own society as a conservative estimate of the SETI capabilities of an extraterrestrial civilization, then the odds of a message from Earth being detected are dismally low simply because of the sheer number of possible frequencies to broadcast on.

This fact has prompted the search for "magic frequencies" that are the most likely to be used as a communication channel. SETI's original magic frequency was 1,420 MHz, which is emitted by hydrogen atoms and considered particularly auspicious because hydrogen is the most abundant element in the universe (Cocconi and Morrison 1959). Moving up the spectrum, one finds the four emission frequencies of hydroxyl (OH) molecules, which range from 1.612 GHz to 1.720 GHz. The range of frequencies spanning from 1.420 GHz and 1.720 GHz is known as "the water hole," given that the combination of a hydrogen atom and a hydroxyl molecule produces water. "Nature has provided us with a rather narrow band in this part of the spectrum that seems especially marked for interstellar contact," NASA concluded in its report on Project Cyclops, the agency's first proposed SETI endeavor. "Standing like the Om and Um on either side of a gate, these two emissions of the dissociation products of water beckon all water-based life to search for its kind at the age-old meeting place of all species: the water hole" (Oliver and Billingham 1972).

Although 1,420 MHz was the chosen frequency for Project Ozma and many other SETI observations thereafter, it may not be as ideal for communication as was once presumed. Frequencies between 1 and 3 GHz (which includes the water hole) have since been shown to be particularly susceptible to interstellar electron clouds that cause a signal's bandwidth to increase. In fact, the most detectable signal that uses the least amount of power was calculated to be at around 70 GHz (Drake and Sobel 1992), which is still within the universe's "quiet zone," but well outside Earth's microwave window. This implies that our atmosphere may be blocking any incoming extraterrestrial signals on this optimal frequency, while also precluding optimized broadcasts. Even if we established a METI outpost on the moon for broadcasting and receiving, the detectability of our signal would now be further restrained by the assumption that our extraterrestrial targets have also overcome their atmospheric opacity. Fortunately, there are other reasonable magic frequencies that still allow for broadcasts from *terra firma* (for a thorough overview of the rationale behind many leading frequency candidates, see Blair and Zadnik 1993 and Townes 1957, 1983). A few notable examples include harmonics of the hydrogen line, at 2.840 GHz; the product of the hydrogen line and pi, justified on the grounds that the irrationality of pi means that this frequency couldn't be produced naturally as a harmonic and would thus distinguish the signal as artificial (Zaitsev 2011); 8.67 GHz, which is the spin-flip transition of 3He+ ion, chosen because it has the next simplest transition after atomic hydrogen (Bania and Rood 1993); and 203.385 GHz, which corresponds "to the splitting of the ground state of the lightest atom—positronium—and [coincides] with the centroid of the relic background spectrum" (Kardashev 1979).

Once a frequency has been selected for interstellar transmission, the next order of business is to determine how information will be encoded in the signal. Modulation is the technique of embedding information by modifying one or more characteristics (amplitude, frequency, or phase) of an analog or digital signal. An analog signal consists of a sinusoidal, or continuous, waveform that can be thought of as representing an infinite number of possible values within a given range. A digital signal, on the other hand, is a stepped waveform that always represents a discrete value. On Earth, anyone who has ever listened to the radio is familiar with the amplitude modulation (AM) and frequency modulation (FM) of an analog signal. While this works well enough for terrestrial communication and is particularly well suited to vocal communication, it has several disadvantages that make it less than ideal for interstellar communication, such as a low signal-to-noise ratio and difficulties with error correction. This does not necessarily preclude the use of analog signals for METI, and analog transmissions do open interesting possibilities for message creation. The 2001 Teen Age Message, for example, intentionally used an analog transmission to indicate that the signal was artistic rather than linguistic, which would manifest as a digital signal (Zaitsev 2008). Interstellar messages have historically relied on digital modulation techniques—known as *keying*—to encode information in an interstellar transmission. Several methods of digital encoding are similar to analog encoding, such as frequency-shift keying (FSK), amplitude-shift keying (ASK), phase-shift keying (PSK), and combinatorial keying schemes such as quadrature amplitude modulation (QAM). Each of these digital modulation techniques modifies a certain characteristic of a signal between two or more discrete values to encode binary information and involves trade-offs in terms of the intelligibility of the signal. FSK, for instance, requires a receiver to monitor at least two frequencies simultaneously, but has a high signal-to-noise ratio. PSK, on the other hand, is less immune to noise, but is far more bandwidth efficient. Simultaneously modulating the polarity of the wave will also help the signal stand out as artificial. The choice of one modulation paradigm over the other depends largely on the parameters of the message and the energy budget for the transmission.

In principle, these modulation techniques are the same whether a signal is sent on Earth or across the cosmos, but interstellar communication introduces several unique difficulties owing to a lack of coordination between the transmitter and receiver. For a naïve illustration of this problem, consider a hypothetical transmission that uses FSK modulation, where a signal alternates between 1,420 MHz and 1,450 MHz to encode 0 and 1, respectively. Let us suppose that the transmitted message consists of 8 bits—00110101—and is transmitted at 1 bit per second. The extraterrestrial will thus receive a signal that consists of a two-second pulse at 1,420 MHz, followed by a two-second pulse at 1,450 MHz, and then four pulses alternating between the two frequencies for one second each. How will the extraterrestrial interpret this signal? The most obvious interpretation is that each frequency represents a binary digit. Yet this introduces a difficulty: the extraterrestrial doesn't know how much data is contained in each radio pulse. For example, a two-second pulse at 1,420 MHz might encode "00" or it might encode "0000." Unless the extraterrestrial knows the bit rate of the transmission, extracting the binary sequence from the signal will be next to impossible.

One method of addressing this issue is to include an additional signal that functions as a clock that the extraterrestrial can use to synchronize its devices to the bit rate of the incoming signal. This clock would essentially consist of an alternating binary stream (e.g., 01010101010 …) where the number of transitions per second corresponds to the bit rate of the data channels. Given the repetitive nature of this signal and its rapid oscillation, it is reasonable to suppose that an extraterrestrial wouldn't mistake it as encoded data, and it could even potentially be used as a beacon that would both draw attention to the signal and synchronize the transmitter and receiver. Another method of calling attention to this signal as a clock is to make the signal carrying the clock significantly more powerful than the data channels. Although a clock could be broadcast as its own signal, it can also be encoded in the message itself. One method of doing this is known as *Manchester coding*, a type of phase-shift keying that combines a clock and the data into a single signal. This is accomplished by encoding binary values as the transition between two discrete states. For example, 0 is encoded as the transition from a "High" state to a "Low" state and 1 is encoded as the transition from a "Low" to a "High" state while each transition represents one cycle of the clock. The trade-off with this method of encoding, of course, is that it doubles the bandwidth required for the signal.

The design of an interstellar radio message thus requires a cost-benefit analysis that considers the bandwidth of the signal, the target star system, the power of the transmitter, the selected frequency, and the data rate of the transmission, as well as some assumptions about the nature of the extraterrestrial receiving technology. Taken together, these set restrictions on the size of the message that is sent. Consider, for example, a situation in which the Arecibo telescope is used to transmit a message at 1,420 MHz to an extraterrestrial civilization 100 light years distant that has a comparable receiving telescope. Best practice in radio communication suggests that the bandwidth of a signal will be between 0.1 and 10 percent of the carrier frequency (Shostak 1995), so a broadcast at 1420 MHz would use a bandwidth of about 70 MHz. If the desired signal-to-noise ratio is 1—a remarkably clear signal—then the equation for the amount of information that can be transmitted over this channel returns 70 megabits per second, or about 750 gigabytes of information each day (Shannon 1948). The feasibility of this scenario depends on the ability to achieve a signal-to-noise ratio of 1, which is a function of the power density of the transmitter. In this case, achieving this signal-to-noise ratio would require a power density of 1 kilowatt per hertz, or 70 gigawatts spread across the entire 70 MHz band. This energy requirement is considerable: it represents approximately 0.5 percent of the total energy generation capacity of Earth, which is far beyond the reach of our most powerful radio telescopes (Shostak 2009). Allotting this much energy solely for interstellar messaging is unrealistic, but the power requirements can be greatly reduced by altering the parameters of the transmission. If we assume that the extraterrestrials' SETI program is comparable to our own, then they will be looking for signals with bandwidths around 1 Hz and will be using instruments sensitive enough to handle a signal-to-noise ratio well below 1. Thus, the power demand for transmitting can be greatly reduced by significantly limiting the bandwidth of a message or the data rate, which will involve a trade-off in terms of the transmission time. Alternatively, one could assume that an extraterrestrial society will have radio telescopes with larger receiving areas than on Earth (i.e., greater than one square kilometer). For example, the 1999 Cosmic Call from the 70-meter Evpatoria telescope in the Ukraine sent messages with a bandwidth of 48 kHz to stars within 70 light years of Earth using only 150 kW of energy and bit rates ranging from 100 to 2,000 bits per second (Zaitsev and Ignatov 1999). Each of these parameters is several orders of magnitude below the hypothetical situation outlined above. Although this message will be quite weak when it arrives at the target star systems, any extraterrestrials on the receiving end of the message should be able to detect it with receivers only slightly more advanced than the most powerful extant radio telescopes on Earth.

If an extraterrestrial can detect a signal, determine its bit rate to extract a raw binary sequence, and then analyze this sequence to determine its encoding mechanism, the final step will be interpreting the contents of the message. On Earth, several standards have been created to coordinate the transmission of digital data and facilitate information exchange among devices. The modern American Standard Code for Information Interchange (ASCII), for example, maps unique 8-bit sequences to 255 linguistic characters, numbers, and control characters. In the case of interstellar communication, of course, these sorts of agreed-upon standards can't be determined in advance. The extraterrestrial will just receive a large undifferentiated stream of binary information and will have to decipher patterns and relationships between the pieces of data in this signal to determine how symbols or other information is encoded. As we have seen, it is paramount that interstellar messages be designed to make this decoding process possible.

OMETI

Prior to the advent of radio communications, every scheme for contacting extraterrestrials involved some sort of optical apparatus. The Austrian astronomer Joseph von Littrow, for instance, suggested creating massive flaming canals to communicate with the extraterrestrials presumed to exist on Mars. The French astronomer Camille Flammarion suggested creating large arrays of electric lamps for the same reason. Despite the best intentions of these early alien hunters, we now know that even if there were intelligent life on Mars, we would never be able to construct mirrors on Earth large enough to reflect light that could be seen by Martians, much less at interstellar distances (Fleming 1909).

Although it is theoretically possible, albeit wildly impractical, to build mirrors of the necessary magnitude in space (Teller, Wood, and Hyde 1997), the main issue with this proposal for interstellar communication is the nature of the light source itself. The lack of coherence in sunlight would severely limit the strength of a mirrored signal. Yet even if we tried to overcome the divergence of sunlight by using a lens that focused the sunlight at our target to increase the signal's strength—similar to the way a magnifying glass can be used to start a fire on Earth—the emission spectra of sunlight would still make our signal difficult to detect for any extraterrestrial optical astronomers. The reason for this is that sunlight is smeared across the visible portion of the electromagnetic spectrum, a potpourri of frequencies slightly biased toward the blue end of the optical spectrum. For an optical transmission to stand out against the starlight, the signal would have to be concentrated in a narrow bandwidth, powerful enough to stand out against the background noise of our Sun, and easily modulated so that information can be encoded in it.

In 1953, a team of physicists at Columbia University led by Charles Townes created the first maser, a device that produced coherent beams of microwave radiation. Shortly thereafter, Townes and his colleagues conceptualized a similar device, the optical maser, that would produce coherent beams of visible and infrared light. By 1960, the optical maser (now better known as a laser) had made its way from theory to reality, and Townes immediately grasped its implication for interstellar communication. In a paper published just six months before Frank Drake convened the Order of the Dolphin at Green Bank, Townes outlined the foundation for what has since become known as optical SETI, or OSETI (Schwartz and Townes 1961). Although Townes conceded that laser technologies were "still in a rudimentary stage," he calculated that an extraterrestrial civilization about as advanced as our own within about ten light years would be able to generate a laser beam that could be detected on Earth. He also foresaw that laser communications would be "quite practical" for interplanetary communication within our own solar system. Lasers could be wielded for interstellar communication either as pulsed signals or as continuous beacons, although continuous signals are harder to pick out against background starlight. Pulsed signals, on the other hand, last only a few billionths of a second and as such can appear several thousand times brighter than their sender's host star without using an excessive amount of energy. These brief pulses of visible or infrared light would distinguish themselves as artificial because light pulses from astrophysical phenomena generally occur on the microsecond scale. Furthermore, interstellar laser communications would be concentrated an extremely narrow range of frequencies, which would be easily detectable as a concentrated energy spike by a sufficiently high-resolution spectrometer.

Information could be digitally encoded in an optical signal through either wavelength modulation, polarity shifts, or pulse timing (McConnell 2001). Lasers are quasi-monochromatic above a certain pulse duration, so digitally encoding messages with wavelength modulation would require shifting between two or more narrowly spaced frequencies. Polarized lasers have an electric field that oscillates perpendicular to the laser beam, so shifting the orientation of this polarization could also be used to encode information. Finally, pulse timing uses short and long laser bursts to encode information in a manner similar to Morse code. Importantly, each of these three methods transmits data at different rates, and all three forms of modulation can be combined in a single pulse to allow more data to be sent or for added redundancy in the message.

The scattering effects of the interstellar medium suggest that infrared lasers are preferable to visible light for interstellar messages sent over distances greater than 100 light years. Given the effects of dispersion on infrared signals (i.e., those with wavelengths greater than or equal to 1 micrometer), however, the distance to the target puts a limit on the number of pulses that can be sent per second. For example, an optical signal with a wavelength of 1 micrometer would be limited to about 1 trillion pulses per second whereas an optical signal with a wavelength of 10 micrometers would be capped at 10 billion pulses per second. Despite these limits, optical interstellar communication is valuable because it is a very information-dense medium compared to microwave communication (Ross and Curran 1965). In a naïve hypothetical scenario where an infrared system is used to transmit one pulse per second and each pulse contains 1 bit of data, it is possible to signal a star system 100 light years away with a photon flux ten times greater than our Sun using only 1 kilowatt of power. Of course, as data rates increase, so do the requisite energy requirements. Compared with microwave signals, optical communication has a far higher energy requirement. For instance, if a wideband radio message used all of the 9 GHz of bandwidth in the Earth's microwave window, this would have a bit rate comparable to the peak data rate of a message sent using infrared at a wavelength of 10 micrometers, but the radio message would require up to eight orders of magnitude less power than the optical message. This example is mostly to demonstrate the widely different power requirements between microwave and optical messaging, rather than to suggest that both messaging techniques are equally feasible. Not only would the encoding scheme necessary to create a message that spans 9 GHz be remarkably complex, it would also require a radio telescope with an effective aperture equal to about half the Earth's diameter to broadcast at the lower end of the microwave spectrum to star systems 1,000 light years away (Shostak 2011).

The first OSETI observation took place in the Soviet Union in 1973 and scanned a handful of stellar targets. Since then, thousands of OSETI observations have taken place and several dedicated OSETI observatories have been established. OSETI seems especially promising because we already have the optical technology on Earth to create sophisticated transmission devices that could beam high powered signals to distant stars, which suggests that any extraterrestrial intelligence in our galactic neighborhood likely does, too. Despite the promise of OSETI, however, there haven't been any "OMETI" projects yet. In fact, NASA only recently began experimenting with laser communications between Earth and the Moon, which is the farthest laser communication link ever established (Boroson et al. 2009). The lack of OMETI projects has more to do with resource conservation than technological limitations, however. Not only are laser transmissions far more energy intensive than microwave transmissions, but optical telescopes also require sophisticated equipment adaptations to make them capable of transmitting laser signals. The only project that is seriously considering a mission that might be classified as OMETI is the Breakthrough Initiative, a research program bankrolled by the Russian billionaire

Yuri Milner to the tune of $100 million. In addition to solic-
iting ideas for interstellar message content (Breakthrough Message) and funding a SETI project (Breakthrough Listen), Milner's
initiative includes Breakthrough Starshot, an ambitious plan
to use a kilometer-scale array of lasers to produce 100-gigawatt
laser pulses that will propel thumbnail-sized nano-craft to Alpha
Centauri at 20 percent the speed of light (Billings 2016). Alpha
Centauri is a tri-solar system that was at one time believed to
host an Earth-sized planet (Dumusque et al. 2012), although this
claim has since been cast in serious doubt (Rajpaul, Aigrain, and
Roberts 2015). Breakthrough Starshot's mission isn't to scout to
for intelligent life around Alpha Centauri, but rather to characterize our closest stellar neighbor. Nevertheless, if the project comes
to fruition it may also qualify as the first OMETI project, given
that leakage from the beamed radiation used to propel the spacecraft will be detectable at interstellar distances (Guillochon and
Loeb 2015).

8

ART AS A UNIVERSAL LANGUAGE

Following the launch of the Voyager golden records, the visual and musical arts came to occupy an increasingly prominent position in the design of interstellar messages. The first instance of art broadcast into the cosmos was the sound of vaginal contractions in ballerinas transmitted from MIT's Millstone Radar by Joe Davis in 1985. Davis's transmission to Tau Ceti and Epsilon Eridani was ultimately cut short by the Air Force after it learned the content of the transmission. In the thirty years since Davis's transmission, however, at least five other artistic messages have been broadcast into the cosmos. The emphasis on artistic elements in interstellar messages is hardly surprising given that every culture on Earth produces visual and auditory art. Indeed, art is often described as a universal language and this suggests that it may serve as a solid foundation for communicating with extraterrestrial intelligences. The apparent advantages of using pictures and music for interstellar communication include information density, reduced abstraction (at least in the case of icons), and unique insight into human cognition. Yet for all their advantages, art-based messages make a large presumption that the extraterrestrial receiver also has an artistic impulse. Even if extraterrestrials do produce their own art, however, this does little to address the far greater problem of how they are supposed to recognize that the content of our message is artistic, rather than linguistic or scientific. For an appreciation of the difficulty of this problem, one need only consider how contentious the definition of art is on Earth. Although most humans have an intuitive sense of art—they "know it when they see it"—there is no universally accepted formal definition of art that differentiates it from other domains. For example, many definitions of art evoke context as a determining factor. This emphasis on context was exploited by Marcel Duchamp's readymades, the most famous of which is *Fountain*. By placing a urinal in a gallery and dignifying it with an exhibition, Duchamp managed to turn an entirely mundane item into a work of art. Thus, a main problem associated with using art in interstellar messages is providing sufficient context for its correct interpretation. Although this problem also haunts linguistic and mathematical messages, it is exacerbated in the case of art vis-à-vis other symbolic communication systems. After all, what are extraterrestrials to make of a painting by Salvador Dalí? Would they presume that Earth is inhabited by melting clocks and elephants on stilts? On Earth, a Bach cantata or Dalí painting elicit emotions and can be "understood" in the context of their milieu, but they defy definite meaning in the sense we understand "Stop!" as a command to arrest our motion. On the other hand, these artworks can be understood on a technical level. Bach's fugues, for instance, are renowned for their mathematical precision, a feature that would likely be appreciated by an extraterrestrial intelligence—but would it feel the melancholy of *St. Matthew's Passion*? One possible solution is to use solely abstract art, which may be considered universally intelligible given its rejection of "cultural, historical, or political contexts" (Brinkmann et al. 2014). This question about what type of art to include in an interstellar message is an important one and well worth consideration. Given the lack of any non-contentious definition of art, however, we will here consider the virtues of two artistic media—pictures and sound—that may be called art only in the loosest sense of the word.

THE CONVENTIONALITY OF IMAGES

Two years after the Voyager spacecraft departed on their journey beyond the solar system, the United States Congress authorized the Department of Energy to build the Waste Isolation Pilot Plant (WIPP) near Carlsbad, New Mexico, "to demonstrate the safe disposal of radioactive wastes resulting from the defense activities and programs of the United States" (Trauth, Hora, and Guzowski 1993). The transuranic materials that produce nuclear waste often have extremely long half-lives on the order of thousands of years, which presents a particularly challenging engineering problem for those tasked with designing storage facilities for spent fissile material. For this reason, WIPP and the handful of other existing nuclear waste repositories are essentially designed like tombs. Once they are closed, they are not meant to be reopened. Yet as the archaeological conquest of Egypt and the sites of other ancient civilizations remind us, humans have few qualms about cracking ancient seals. There is little reason to suspect that future humans will not be driven by a similar curiosity. Although disturbing Tutankhamun didn't bestow a curse on Howard Carter's excavation crew, any future Earthlings who dared enter our nuclear waste repositories would not be so fortunate.

The importance of protecting WIPP, the first dedicated subterranean nuclear waste repository in the United States, from future human intrusion wasn't lost on its architects. In 1985, the US Environmental Protection Agency issued a directive that highlighted the need for warning markers that would prevent inadvertent human intrusion for ten thousand years—the regulatory lifetime of the repository. Considering that the oldest written records on Earth only date back about five thousand years and the meaning of several ancient scripts has been lost to history, creating a message that would be intelligible for twice that time period was a daunting challenge. In accordance with the EPA's directive, Sandia National Laboratories convened a working group of scientists, linguists, anthropologists, and artists who were tasked with designing warning messages that would be intelligible for ten millennia. A notable feature of the thirteen-member task force was the strong representation of SETI researchers, including Woodruff T. Sullivan, Frank Drake, and Jon Lomberg, the artist who designed the original iconography for the Voyager golden records.

In a letter declining an invitation to participate on the task force, Carl Sagan eloquently framed the problem at hand. "Social institutions, artistic conventions, written and spoken language, scientific knowledge, and even the dedication to reason and truth might, for all we know, change drastically," Sagan wrote. "What we need is a symbol invariant to all those possible changes. Moreover, we want a symbol that will be understandable not just to the most educated and scientifically literate members of the population, but to anyone who might come upon this repository." Sagan's own recommendation for a ten-thousand-year warning message was the skull and crossbones, an emblem that seemed to have an "unmistakable" meaning. "It is the symbol used on the lintels of cannibal dwellings, the flags of pirates, the insignia of SS divisions and motorcycle gangs, the labels of bottles of poisons," Sagan reasoned. "Human skeletal anatomy, we can be reasonably sure, will not unrecognizably change in the next few tens of thousands of years" (Trauth, Hora, and Guzowski 1993). Sagan's conviction about the universality of the skull and crossbones as a warning emblem is intuitive, but unconvincing. This macabre iconography is a distinctly Western symbol that likely emerged from masonic culture and, while always a symbol of death, only came to globally signify toxicity or danger relatively recently. Thus, if the futurists are right and we are on the cusp of the integration of human and machine, our cyborg ancestors may interpret the skull and crossbones at the entrance to our nuclear repositories as the marking of an ancient, twentieth-century ossuary filled with the remains of their ancestors who were still more flesh than silicon. One can hardly imagine a more titillating prospect for the archaeologists of the future.

The point is that all images are burdened with ambiguity, which is one of the many reasons that the designers of the WIPP message cautioned against relying too heavily on artistic warning messages. Historically, image-based messages have been embraced as a superior mode of interstellar communication in terms of both information density and the apparently direct connection between the image and the thing it represents. The term for an image that bears a physical resemblance to the thing it represents is an "icon," whereas an image that has an arbitrary, conventional connection to the thing it represents is a "symbol." For example, Gauss's proposal for an interplanetary message made of a massive pictorial proof of the Pythagorean theorem in the Siberian tundra is iconic, whereas "$a^2 + b^2 = c^2$" is symbolic. Yet as we saw in the case of using a skull as the WIPP warning message, whether an image is interpreted as a symbol or an icon can still be a matter of convention and context.

It has been argued that icons are preferred for interstellar messages because they drastically reduce the amount of interpretive difficulties associated with symbols (Vakoch 1998b). Still, it seems that a great deal of convention is bound up even in icons, which depend on subjective or culture specific models of reality (Lemarchand and Lomberg 2009). Consider, for example, the story of an encounter between a Swiss artist and a member of the Sioux tribe in the nineteenth century. When the Swiss artist drew a profile of a man on horseback, only one leg could be seen from the perspective of the artist and the other was hidden by the horse. Yet when the Sioux drew the same scene, the man on the horse was in profile, but both legs were visible to the viewer. Although the Swiss artist protested the unreality of the indigenous representation, the Sioux had a ready defense: "But, you see, a man has *two* legs" (Highwater 1983). In this case, both the Swiss and Sioux artist were representing the same objective reality, but their models of reality were different. This is analogous to the incommensurability problem that faces scientific interstellar messages, which presumes that extraterrestrial science will be the same as our own. Similar difficulties are inherent to imaged-based interstellar messages. Consider Drake's "pictogram" sent from Arecibo in 1974. The bitmap contains an iconic representation of DNA, but there is little reason to think that this two-dimensional representation of DNA's helical structure is the only way of representing it or that it would be interpreted as such. Indeed, when the neuroscientist Michael Arbib examined Drake's 1960 prototype interstellar message, he demonstrated how Drake's representation of atoms could be interpreted as "six-legged ... large-brained creatures with tails" (Arbib 1979). This problem is not unique to bitmaps but also holds for the photos contained on the Voyager craft and encoded in the Cosmic Call messages. Consider, for example, if we wished to send a photo of the Earth to extraterrestrials. Should we send a time lapse that conveys the planet's rotational velocity, or a snapshot of an Earthrise from the surface of the Moon? What about a composite of the planet that removes all the cloud cover? None of these images is an "objective" representation of the Earth. Each contains the bias of the photographer, who has made a conscious decision to emphasize some aspect of the planet over others in the image.

A final issue with visual messages is their inability to convey statements. Although images are information rich, they are severely limited in their ability to communicate statements about abstract ideas. For instance, consider the difficulty of visually representing "This is a statement" or its negation. Yet, as the art historian Ernst Gombrich (1972) noted, the poverty of images in communication is not solely due to the abstractness of natural language. If one were to visually represent a statement such as "the cat is on the mat," one might draw a cat sitting on a mat. This fails to capture important aspects of the statement, however, such as whether we are talking about this individual cat or this cat as a representative of the class "cats," whether this statement is actually about a cat sitting on a mat as opposed to demonstrating what a cat looks like from behind, or whether this cat is currently sitting on the mat, sat on the mat in the past, or will sit on the mat in the future. As Arbib later summarized the issue, "the trouble with pictures is that they are too literal to communicate general truths" (1979).

In our increasingly image-saturated world, to send an inter-stellar message devoid of any visual components would be to leave out a large part of what it means to a human on Earth in the twenty-first century. Although they are highly conven-tional, pictorial messages are not necessarily a lost cause when it comes to interstellar communication, so long as they are accompanied with ancillary information. One way to go about this is by providing contextual information that can introduce unfamiliar objects in terms of familiar mathematical concepts. Consider the Voyager records, which used the spin-flip transi-tion of neutral hydrogen and corresponding emission frequency and wavelength as a basic unit of measurement for time and distance to provide a scale for the photos. Alternatively, Ollon-gren's Lincos could also be used to describe pictures in terms of a metalanguage based on constructive logic. Another option for disambiguating images is by rendering them in three or four dimensions. This would eliminate the need to "represent objects from a single, privileged vantage point," but there would still be aspects of conventionality in these (moving) images, such as the frame rate and what is chosen for representation. For example, a four-dimensional image depicting a human walking could suggest to an extraterrestrial that "we deem ourselves to be an important topic of communication" (Vakoch 2000b).

Four-dimensional images or "moving pictures" are an espe-cially attractive medium for communication insofar as they also capture the all-too-human impulse to tell stories and link individual events into cohesive narratives. To this end, Harry Letaw Jr. (2005, 2013) proposed a cinematic messaging scheme in which moving pictures are supplemented with a symbolic system. Nevertheless, messages relying on moving images must still address an issue noted by Sol Worth in his seminal study on visual communication, where he stated that "an objective, value-free film record" is impossible since "every film maker has an inherent cultural bias" (Worth 1981). This observation was drawn from Worth's ethnographic research on the Navajo Nation, during which he showed a Navajo audience a selection of films that were made by non-Navajo filmmakers. These films didn't include a sound track and were all in English, which led one Navajo viewer to remark that they couldn't understand the meaning of the film because of their lack of familiarity with the language. The deeper implication here is that the visual metaphors in the film were not enough to convey their meaning to the audience without the assistance of language, owing to the cultural-specificity of the metaphors. To address this issue in an interstellar context, Letaw proposed using "Rosetta elements" that would consist of visual representations that are stripped of as much cultural metaphor as possible. These Rosetta elements would consist of actions that might be familiar to an extraterrestrial and include such classes as physical observables (weather events, wave motion, light and shadow, etc.), simple machines (levers, wheels, hammers), basic physical activities (walking, running, eating), and perhaps even domestic pursuits (erecting shelters, hunting, agriculture). Thus, if our extraterrestrial recipient is supposed to live on a planet that has large bodies of liquid and a weather cycle, a message might include a high-speed film depicting a water droplet falling into a body of water. This "movie" might then be overlaid with a symbolic system describing what is happening in mathematical terms. Based on these Rosetta elements, Letaw suggests it would be possible to craft increasingly sophisticated narrative films for our interstellar messages in order to depict abstract concepts such as human altruism.

Although this cinematic approach allows for incredible diversity in message construction and resolves many of the ambiguities inherent in two-dimensional still images, it is haunted by the problem of selecting primitives. As Leibniz discovered, a communication system based on a collection of primitive elements is not a suitable basis for a universally intelligible language because they are arbitrary. Although it is reasonable to assume that an extraterrestrial might be familiar with the concept of a weather cycle, the same cannot be said of even the most basic of human activities such as sleeping or shelter construction. Although Letaw had no illusions that we can "assure understanding on the part of the ETI" when sending narrative films as part of an interstellar message, the persistent ambiguities inherent in visual communication suggest that it would be worthwhile to consider other avenues of message construction.

MUSIC OF THE SPHERES

As a system whose features include melody, rhythm, pitch, and harmony, music has been identified in every culture on Earth, even if the forms this music takes are quite different. Music appears to activate many of the same regions of the brain as language and mathematics, but it is far from certain that musical ability is innate (Patel 2003). In fact, music seems more like a "cultural technology" that has been developed over thousands of years. While every human develops language except in cases of extreme pathology, most humans can appreciate music even if they cannot produce it. Furthermore, although *some* form of music exists in every culture, ethnomusicologists haven't established a consensus on what features of music are truly universal.

Despite the uncertainties about music's status as a universal, it has historically occupied a prominent place in interstellar messaging. This began with the Voyagers' golden records, which included a variety of folk songs from different cultures in addition to songs representing more recent musical innovations such as rock and roll. Embedded in the 2001 Teen Age Message broadcast from the Evpatoria radar in Ukraine was a live theremin concert orchestrated by Russian teens who selected seven songs to be performed by musicians from the Moscow Theremin Center. Shortly thereafter, Alexander Zaitsev orchestrated the second Cosmic Call message in 2003, which included an album by the Hungarian rock band KFT. In 2008, the National Aeronautics and Space Administration broadcast the Beatles' "Across the Universe" toward Polaris to commemorate the anniversary of the song's recording. More recently, the newly minted nonprofit METI International sent two messages from the EISCAT radar in Norway, each of which included several short electronic music pieces produced by musicians affiliated with the Sónar music festival.

Of these musical transmissions, only the Teen Age Message and the two Sónar messages orchestrated by METI International are scientific in their design. Although NASA's "Across the Universe" message was widely (and erroneously) heralded in the press as the first instance that music was used for an interstellar broadcast, it is almost certain that any extraterrestrials that may be on the receiving end of the transmission will be unable to understand it. In the first place, the coding and compression standards used to digitize music on Earth were not accounted for, so it is unlikely that an extraterrestrial that is totally unfamiliar with Earth technologies would be able to extract the content of that message. Even if this problem had been addressed, however, the transmission rate of the message—128 kbps—is roughly 300,000 times faster than the minimum transmission rate needed for an extraterrestrial to extract a message from the signal using receivers comparable to those on Earth. If this message were to be rebroadcast at an "admissible" rate closer to the minimum, that would greatly improve the likelihood that it would be deciphered by an extraterrestrial recipient, but it would've taken nearly 750 days of continuous broadcasting to send a single Beatles song (Zaitsev 2008). In the case of the Teen Age Message, Zaitsev abandoned the usual approach of embedding messages in a binary signal and opted instead to send music directly as a continuous quasi-sinusoidal signal. To date, the Teen Age Message is the only interstellar broadcast sent as a continuous signal rather than a discrete signal that shifts between two or more frequencies to encode information. This design choice was motivated by the musical content of the message: a continuous signal helps distinguish the message not only from astrophysical sources, which produce a constant frequency function, but also from artificial messages carrying logical or linguistic information, which would manifest as a frequency shift between two or more discrete values. In short, a continuous quasi-sinusoidal signal has the apparent benefit of being able to distinguish the signal as both artificial *and* artistic in nature (Zaitsev 2008).

A further benefit of an analog interstellar signal over digital methods can be seen in the drastically reduced transmission times. Consider a hypothetical transmission scenario that involves transmitting the signal to an extraterrestrial civilization seventy light years away that possesses a radio telescope whose receiving area is equal to one square kilometer and has a noise temperature slightly better than has been developed on

Earth. In this case, transmitting the theremin concert portion of the Teen Age message would take only fourteen minutes using analog encoding as opposed to nearly fifty hours of transmission time for the equivalent message encoded digitally (Zaitsev 2008). Theremins were chosen for the interstellar broadcast because they produce sustained (self-oscillating) signals that don't fade over time, unlike the damped oscillations of drums, pianos, or guitars. A related advantage is that the oscillations produced by the theremin are concentrated in a narrow band around the fundamental tone, minimizing the level of overtones and allowing for smooth transitions from one note (frequency) to another instead of a stepped transition. Together, these serve to concentrate the theremin signal into a narrow range of frequencies, which increases the chance that it will be detected by a receiver against background noise. From a technical perspective, theremins are also easily integrated into a radio-transmission system. These instruments work by down-converting the high-frequency oscillator that is the core of the theremin into the audible spectrum by mixing the high-frequency oscillations with the frequency of a reference oscillator. To convert this into an interstellar radio signal, an opposite transformation occurs, and the high-frequency oscillations are up-converted into microwave frequencies to be broadcast into space. The theremin has an added benefit that that its high-frequency oscillations can be converted into any band of the electromagnetic spectrum. Thus, if an extraterrestrial cannot hear sound in the same frequency as humans, it is possible to convert the signal into optical wavelength to produce "color music" that is purely visual. Finally, theremins have historically occupied a prominent place in the soundtracks of science fiction films dealing with extraterrestrial life, so their use for an actual interstellar message only seems fitting.

As for the Sónar messages, these transmissions were notable for being the first messages broadcast toward a known exoplanet. The destination, known as Luyten's Star, is a red dwarf about 12.4 light years from Earth. In 2017, two planets were discovered orbiting Luyten's Star; one of them, GJ237b, is a planet nearly three times the mass of the Earth orbiting in the star's habitable zone. The first Sónar message consisted of thirty-three brief musical compositions of about ten seconds each that were preceded by an initial "hello" beacon and a tutorial for interpreting the message. The purpose of the beacon is to draw the attention of any extraterrestrials that might be listening on GJ237b by differentiating the signal from natural radio sources. This was accomplished by encoding a series of primes (from 1 to 137) in two different radio frequencies and broadcasting each prime at each frequency before proceeding to the next prime. Each number was represented by an equal number of pulses at the specific frequency with a 16 millisecond pause between consecutive pulses. After 137 is reached, the message starts again, but twice as fast and counts the primes to 193. The beacon ends by introducing the concept of "bytes," which are 8-bit blocks of binary information. This part of the beacon counts from 0 to 255, the maximum number that can be encoded in 8 bits: 00000000, 00000001, 00000010 ... 11111111.

The next part of the Sónar message is a tutorial, which differed slightly in the first message and the second message. In the first message, the tutorial used bytes to introduce a small collection of concepts, such as arithmetic operators and physical phenomena such as the concept of sound, frequencies, and the speed of light. The tutorial also used bytes to describe simple sounds as a sine wave, which is the same format of the musical pieces themselves. Thus, this part of the tutorial begins with a 100 hertz wave before moving to 200 hertz, 400 hertz, 800 hertz, and 1,600 hertz. Finally, it demonstrates how these different frequencies can be combined to produce more complex sounds. The second Sónar transmission, sent in early 2018, has the same three-part structure—beacon, tutorial, music—but the contents of its tutorial are quite different. The second tutorial was designed by Yvan Dutil, who also codesigned the message for the Cosmic Calls in 1999 and 2003. For the Sónar message, Dutil drew heavily on the conceptual framework of the Cosmic Call messages to create a self-interpreting message based on a dictionary of concepts encoded as small images. The final tutorial is a bitmap that depicts concepts such as arithmetical operators, chemical formulas, and human anatomy. The final part of the second tutorial was designed by researchers from the Institute of Space Studies of Catalonia, who graphically demonstrated the concept of bytes by counting from 0 to 255 and representing each of these numbers as a group of pixels in a bitmap and then graphically representing a short sequence of bytes as a sinusoidal wave.

The actual musical samples sent along with the Sónar messages included an AI-generated "musical language," the sonification of the binary representation of pi, field recordings, and electronic music. Of course, there is a very real possibility that any extraterrestrial intelligence on GJ237b is incapable of hearing or hears at a different range than that of humans, who are limited to frequencies between 20 Hz and 20,000 Hz. Even so, these musical messages can still teach an ETI quite a bit about human cognition (Kaiser 2004). Humans treasure music for its emotional and aesthetic qualities, but its beauty can also be seen when translated into mathematics and physics, a property that makes it an attractive medium for interstellar communication. The mathematization of music can be traced to Pythagoras, the Greek philosopher and mystic who was the first to codify the relationship between the pitch of a note and the length of a string that produced it. Interestingly enough, Pythagoras and his acolytes extended this concept of musical harmony to the motion of planetary bodies as the so-called music of the spheres. As Aristotle recounted this system, the Pythagoreans saw that "all things seemed to be modeled on numbers, and numbers seemed to be the first things in the whole of nature, they supposed the elements of number to be the elements of all things, and the whole heave to be a musical scale and a number" (Aristotle 1912). Recent research has demonstrated that planetary harmonics actually exist, although not as a result of any metaphysical law or a universe designed with an eye toward mathematical elegance. Rather, the orbital periods of planets in our solar system and a few others have been shown to be related by ratios of small integers, which results from periodic gravitational influence among the planets, a phenomenon known as orbital resonance. The important point, however, is that the development of musical theory on Earth is deeply related to our understanding of mathematics, and as we saw in earlier chapters, our mathematical reasoning is tied to our embodied cognition.

At a basic level, a musical message can demonstrate the physiology of human sonic perception. Humans can hear tones ranging in frequency from about 20 Hz to 20,000 Hz, but most musical instruments only operate within a limited portion of this spectrum—about 50 Hz to 5,000 Hz. Thus, if an instrumental message is sent to an extraterrestrial, they might mistakenly believe that the range of our instruments is the range of our hearing. This suggests that auditory messages could stand to benefit from including the full range of the frequency spectrum in the message, linearly progressing from 20 Hz to 20,000 Hz. This would also provide insight into the physiology of human hearing insofar as our frequency range is limited by the response rate of our nervous system (Gold and Pumphrey 1948). Similarly, we might experiment with dynamic range in our interstellar messages to demonstrate that we are able to register sound intensity over 12 orders of magnitude. As for cognition, a sufficiently rich corpus of terrestrial music included in an interstellar message would demonstrate to an extraterrestrial our preference for certain structures and patterns in musical arrangements. Western classical music, for example, is dominated by a 12-tone equal temperament scale, a system of tuning where the ratio between the frequencies of adjacent notes is the same throughout the scale. The 12-tone temperament results in five harmonics, but there are several other possible foundations for equal temperament scales in polyphonic music. The Arab tone system, for instance, is based on a 24-tone equal temperament scale (Touma 1996). Each equal temperament tuning system involves a tradeoff between the divisibility of an octave into equal parts and the frequency ratios, which ultimately dictate the number of available harmonics within the scale. The choice of temperament in the music samples may provide insight into human music perception insofar as it reveals our sensitivity to tonal changes. A musically inclined extraterrestrial with a more sensitive auditory apparatus might prefer a 53-tone or even 72-tone scale, and these differences in musical sensibility could reveal much about the transmitting civilization (Vakoch 2010). Providing a wide variety of musical samples may also allow extraterrestrials to extract more general features of human musical patterns by studying the physical characteristics of the sound. On Earth, we analyze music in terms of pitch, tempo, and other stylistic elements, each of which can be translated into a physical characteristic of the sound wave. Spectrum analysis could reveal musical overtones and could thus be considered a stand in for pitch; analysis of motion could replace harmonic progression by revealing patterns of chords over time; and pulse could stand in for meter. Analyzing a music corpus through the physicality of the sounds can thus reveal general patterns across styles and cultures (Kaiser 2004).

An important question when it comes to musical interstellar messages is how to convey the musical contents. One option is to send the music directly as a continuous analog wave, as was done for the theremin concert portion of the 2001 Teen Age Message, or as a physical artifact, as was the case with the Voyager records. These direct approaches take for granted that the music will be intelligible to an extraterrestrial on the assumption that music exhibits certain structures that may be considered universal. A more skeptical approach abandons this assumption and assumes that the way humans experience features of music such as pitch or tempo will need to be taught to the recipient. One method of doing this is to take an iconic approach, where the physical qualities of the radio wave are used to convey musical concepts (Vakoch 2010). Although the transmission frequencies used for interstellar communications are well beyond the range of human perception, they share the same fundamental characteristics of audible sound waves. For example, an interstellar signal could be modulated rhythmically as an iconic representation of our notion of rhythm (Vakoch 2004b). A musical message may also help an extraterrestrial interpret the logical or linguistic features of a symbolic message. Ollongren (2004) demonstrated how Indonesian gamelan music could be used to elucidate the logical structure of the Lincos metalanguage that encodes the linguistic contents of an interstellar message. So even if we don't presume that an extraterrestrial has any music of its own, the inclusion of a musical component in interstellar messages can still be fruitfully employed as a mechanism for teaching extraterrestrials about human cognition and physiology, as well as a primer for interpreting the nonmusical components of a transmission.

9

THE MANY FUTURES OF METI

METI has been enveloped in controversy from the start. In 1974, Frank Drake sent the first message across the cosmos in front of a captivated audience of some 250 people at the Arecibo Observatory. Shortly thereafter, Sir Martin Ryle, the Astronomer Royal of England, sent "a letter full of outrage and anxiety" to the president of the International Astronautical Union asking that the organization formally ban the practice of interstellar messaging (Drake and Sobel 1992). In a subsequent letter to Drake, Lyle reiterated his qualms and concluded that it was "very hazardous to reveal our existence and location to the Galaxy; for all we know, any creatures out there might be malevolent—or hungry." After discussing the matter with Drake, the IAU didn't deem the hazard of broadcasting great enough to warrant any action. Nevertheless, the anxiety about messaging extraterrestrials hasn't subsided. If anything, the calls for a moratorium on METI have intensified as the technological capacity for interstellar messaging has improved (Musso 2012). In 2015, over two dozen scientists, academics, and industry leaders affiliated with the University of California Berkeley's SETI program, arguably the leading search effort in the world, signed a statement calling for a moratorium on interstellar broadcasts until "a worldwide sci-entific, political, and humanitarian discussion" occurred. These critics of METI raised four principle arguments against transmis-sion, which can be characterized as "shouting in a jungle," the pseudoscience argument, the profligate transmissions argument, and "Who speaks for Earth?"

SHOUTING IN A JUNGLE

The "shouting in a jungle" critique is inherited from Ryle. It is based on the historical observation that contact between cul-tures with asymmetrical levels of technological development has frequently led to the extermination of the less technologi-cally advanced culture. Considering that any extraterrestrial intelligence we may contact is likely to be more technologically advanced than our own, Earth's history suggests that contact with ETI may result in human annihilation. Some versions of this argument envision a fleet of intergalactic warships using wormholes to physically traverse the vast emptiness of space, while others suggest that extraterrestrials could wreak havoc from a distance by responding to one of our broadcasts with a message containing an otherworldly computer virus (Hippke and Learned 2018) or a projectile launched at a small fraction of the speed of light (Brin 2014). Thus, the argument goes that broadcasting our existence to an unknown other would be akin to "shouting in a jungle" and waiting for the beasts to arrive (Shuch and Almar 2007). After all, we've noticed a conspicuous lack of chatter in the galaxy after well over one hundred SETI searches spanning half a century. Is it possible that advanced ETIs know of a danger of which we are ignorant and are keeping quiet to survive?

Until first contact, discussion about the likelihood that an extraterrestrial will be altruistic or malevolent amounts to pure conjecture. Although plenty of ink has been spilled crafting reasoned arguments for both positions, this discourse ultimately reflects the disposition of the authors rather than knowledge about the universe. In any case, if calls for a moratorium on interstellar broadcasts are grounded in safety concerns for our planet, it may already be too late. In the years since the Second World War, there has been an explosion in the amount of high-power radio transmissions on Earth from television and radio broadcasts as well as astronomical and military radar systems. All sufficiently powerful radio activity above about 10 MHz results in leakage into space, but the nature of the leakage depends on its source. Radio and television broadcasts, for example, are common all over the planet, so this results in near-isotropic radiation into space. This creates a sort of sphere of radio noise around the Earth whose radius—currently about 80 light years—is equivalent to the amount of time these broadcasts have been occurring. The same is true for the hundreds of radar installations that dot the Earth's surface. These radars, such as the Arecibo Observatory and Evpatoria telescope, as well as the numerous military systems, use powerful, focused radio beams to track asteroids and missiles, which radiate into space "like pins on a pincushion, with most of the beams concentrated in the northern hemisphere" (Haqq-Misra et al. 2013). Although these beams are narrow and cover only a small portion of the sky in aggregate, the Earth's rotation leads to a broad sweeping pattern that increases the amount of sky exposed to these powerful radar signals (Ekers et al. 2002).

In theory, any extraterrestrial civilization with a sufficiently sensitive radio telescope would be able to detect Earth's radio leakage, thus blowing Earth's cover. Although it is certainly a "sobering thought that the only signs of intelligent life on Earth which are detectable over interplanetary distances are housewives' daytime serials, the rock-and-roll end of the AM broadcast band, and the semi-paranoid defense networks" (Sagan 1973), this also raises the question of just how sensitive an extraterrestrial's receiver would have to be to pick up Earth's leakage in the first place (Sullivan, Brown, and Wetherill 1978). Given that the strength of a radio signal decreases with distance according to the inverse square law, the power of the signal at its source largely determines the quality of the receiver needed to detect it at a given distance from Earth. If we assume that extraterrestrial receivers are no more advanced than our own, they would have to be prohibitively large to detect radio leakage from television broadcasts. Furthermore, detecting television broadcasts would require signal integration periods on the order of several months, which is impossible given the Earth's rotation, but it's unlikely the signal would even be coherent by the time it reached an ETI since frequency and phase differences introduced government regulatory bodies would cause the signals to cancel one another at large distances (Billingham and Benford 2011). Cold War–era radars, however, produced relatively narrowband radio leakage that is orders of magnitude more powerful than television broadcasts and would thus be much easier for an alien civilization with comparable receiver technology to detect, even at distances of up to several hundred light years (Shostak 2013). The detectability of these radars is also dependent on the path of their beam intersecting with a star system, which is "insignificantly small" (Zaitsev 2007). Indeed, an attempt to quantify this probability found that the odds of an ETI intercepting an asteroid-hunting radar signal from Arecibo is less than 1 in 500,000 (Billingham and Benford 2011).

Given the small probability of detection from radio leakage, critics of METI haven't argued for complete radio silence on Earth. Instead, their concerns have focused on past and future transmissions, which are more likely to be detected and deciphered because they are purposefully directed toward star systems and encoded to encourage the extraction of their content. For some critics, it is the content of the message, rather than the detection of a radio signal, that is the greater danger since the content reveals that intelligent life on Earth consists of carbon-based meatbags and that our technology is inferior to their own. Fears about extraterrestrials extracting the content of past transmissions may be overblown, however. An analysis of the 1999 Cosmic Call message, for example, found that an ETI would be able to recover the information content of the slower portion (100 bits per second) of the transmission at a maximum distance of about 19 light years. This assumes that the ETI has a receiver comparable to the Square Kilometer Array (SKA), which will be the most sensitive radio telescope on the planet when it is completed. The energy of this transmission, however, could be detected at around 650 light years and likely recognized as an artificial signal, but its information content would be lost to noise at this distance and data rate (Billingham and Benford 2011). Increasing the likelihood that an ETI within 100 light years will be able to detect and extract the information content of an interstellar message would require continuous broadcast times on the order of several months or even years. To date, however, most messages have continuously broadcast for only a few hours at most. Indeed, many messages are broadcast for only a few minutes, which suggests they should be "considered as symbolic or demonstrations of human technology rather than serious efforts to converse with extraterrestrial civilizations" (Haqq-Misra et al. 2013).

Short of an outright moratorium on all interstellar communications, some researchers have attempted to outline protocols that could help assess the risk incurred for a given transmission. In 2005, the Hungarian astronomer Iván Almár proposed the San Marino Index, an ordinal ranking of the risk of a given interstellar transmission on a scale from one to ten. The San Marino Index draws its inspiration from the Torino Scale and Rio Scale, which are used to quantify the risk of an asteroid impact and the importance of a candidate SETI signal, respectively. The San Marino Index is determined by adding two variables that represent the intensity of a transmission (I) and the information content of the message (C), represented by the following equation:

$$SMI = I + C$$

Each variable can take on a value from 1 to 5. The intensity of the signal (I) is a logarithmic measure of the strength of the signal relative to the radiation of our own Sun at the frequency and bandwidth of transmission. For example, a signal that was 1,000 times the intensity of the Sun's radiation would have an (I) value of 3. While there is no objective basis for quantifying the impact of a message's contents (C), it is still possible to assign an ordinal value to the term. An interstellar beacon carrying no message content would be ranked lower than a sustained transmission with a multilevel message, for instance. The San Marino Index was adopted by the International Astronautical Association SETI Permanent Committee in 2007 and underscores the notion that "not all transmissions can be considered equal"

(Shuch and Almár 2007). Ideally, the San Marino Index would be used to determine whether a temporary moratorium on a planned broadcast is necessary. A message that scores low on the scale can be presumed to carry little risk, whereas a message high on the scale would likely require international consultation. Beyond the San Marino Index, Billingham and Benford (2011) recommend that all future transmissions must describe the likelihood that the transmission could be detected based on assumptions about the sensitivity of ET receivers and that this description should be subject to rigorous peer review. They argue that this should include a thorough description of the transmitter parameters (e.g., power, frequency, and bandwidth) as well as the message parameters (e.g., bitrate, keying method, and the number of times the message is repeated), and that this information should be made available online in a standardized format.

IS METI SCIENTIFIC?

Another criticism of METI is that it is a form of "unauthorized diplomacy" rather than a hard science (Michaud 2005; Gertz 2016). This critique depends on how science is defined, however, which is a contentious topic among philosophers. One of the most widely accepted delineations between science and pseudoscience was advanced by the philosopher Karl Popper (1959), who defined a scientific hypothesis or theory as one that is falsifiable. As far as METI is concerned, the accusation that interstellar messaging constitutes a pseudoscience has more to do with the transmission of messages than their design. For example, METI practitioners could advance the hypothesis that an extraterrestrial exists in a given star system and send a message to test this hypothesis. However, the lack of a reply to the message does not confirm or refute the existence of an extraterrestrial intelligence in that solar system. Perhaps the message was missed by the extraterrestrial, or they simply decided not to respond. Of course, a similar claim to pseudoscience could be lodged against SETI. In both cases, there are never refutations of the hypothesis, only confirmations.

Although many interstellar transmissions have not been scientific, hypotheses about its design are in principle falsifiable. Consider the hypothesis that extraterrestrials will understand our math. Even if we receive a reply from an extraterrestrial, this is no guarantee that the message was understood, a lesson Frank Drake learned after he received Barney Oliver's reply to his prototypical interstellar message. In that case, Oliver had understood the encoding of the message, but failed to interpret its contents. A similar principle is at work for the design of messages that are broadcast into the cosmos, although their falsifiability depends on a reply. To make these messages scientific, they should include questions for the extraterrestrial to answer in its response to demonstrate understanding, not merely receipt. It is also possible to scientifically test the design of interstellar messages on Earth. The planetary scientist Michael Busch and the physicist Rachel Reddick (2010) tested binary encoding schemes similar to the one proposed by Freudenthal to determine whether this type of message could be decoded if received on Earth. In this test, Busch designed a binary message consisting of about 75 kilobits of information and provided this message to Reddick and five other volunteers who were tasked with interpreting it. The message was designed such that 10 to 20 percent of the bits were randomly missing, to model a situation where late detection and instrument downtime lost portions of the message. The binary message established units of measurement that were then used to describe aspects of our solar system, such as the mass and radius of the Sun and the masses and temperatures of the planets. The highly structured coding scheme allowed Reddick to determine that certain strings corresponded to delimiters such as "=" and arithmetic operators. Reddick was able to correctly interpret the contents of the message with about twelve hours of work, suggesting that it is possible to "establish a common vocabulary and describe the solar system in considerable detail" with a limited amount of information. Although these methods don't guarantee that those messages will be understood by extraterrestrial recipients, they help increase confidence that a transmission will be understood as intended.

PROFLIGATE TRANSMISSIONS

Just a month before the United States dropped nuclear weapons on Hiroshima and Nagasaki, the head of the US Office of Scientific Research and Development that oversaw almost all military research during World War II submitted a report to President Henry Truman that made an impassioned case for the value of fundamental science. The report argued that basic research, or the pursuit of knowledge about the world "without thought of practical ends," was the bedrock of technological innovation and was thus integral to the future prosperity of the United States (Bush 1945). It was a landmark piece of science advocacy that set the course for public expenditures on scientific research in the United States for decades to come. Indeed, the report ultimately resulted in the establishment of the National Science Foundation, which funded the creation of the National

Radio Astronomy Observatory where Drake would inaugurate the modern search for extraterrestrial intelligence in 1960. It's difficult to imagine a more fundamental research paradigm than interstellar communication. Although there is little doubt that first contact would dramatically affect terrestrial religion, politics, and scientific knowledge (Tough 2000), the long odds and ambiguous nature of this event have resulted in SETI programs having to fight tooth and nail to justify their costs over the past half century. Although NASA was involved with a handful of SETI projects in the 1960s and '70s, all federal funding for the search ended in 1993 after the US Congress voted to eliminate funding for the agency's first intensive SETI project, the High Resolution Microwave Survey (Garber 1999). Except for China's recently completed 500-meter Aperture Spherical Telescope (FAST), all SETI programs on Earth for the past quarter-century have been privately funded. Still, the millions of dollars in public funds that have been allocated for SETI dwarf past public spending on METI projects, which has amounted to a few minutes of radar time, a NASA-sponsored monograph (Vakoch 2014), the Pioneer plaques, and the Voyager golden records.

Today, both SETI and METI are entirely dependent on private funding, but this capital is mostly given to SETI projects. This investment disparity underscores a popular criticism of METI that characterizes transmissions as an extravagant waste of material resources that represent "a gamble with high costs and unknown rewards" (Haqq-Misra et al. 2013). To be sure, transmitting messages into the cosmos is vastly more energy intensive than passively listening for an interstellar signal. Whereas SETI is approaching the ability to do all-sky surveys across millions, or even billions, of frequency channels (Garrett, Siemion, and van Cappellen 2016), interstellar messaging requires focusing on a single star and broadcasting on a narrow range of frequencies at a time. Given the billions of possible frequencies and thousands of stellar targets within one hundred light years of Earth, the odds of choosing the correct frequency *and* target are distressingly low. Moreover, METI transmissions have historically lasted only a few hours at most owing to fierce competition for access to radar facilities; making these signals detectable would involve continuous broadcast times on the order of months, if not years. A dedicated optical or radio METI transmitter would resolve this issue, but even then, the energy costs to maintain a sufficiently powerful continuous signal would require an investment of millions of dollars sustained over multiple generations to allow time for a response.

The many unknowns involved with interstellar communication, in terms of both the parameters of a transmission and the results of contact, make it exceedingly difficult to determine if the large expenditures required for continuous transmission will be worth it in the long run. Proponents of interstellar messaging argue that the knowledge to be gained from contact is invaluable, whether this is through exposure to extraterrestrial technology á la *Contact* (Sagan 1985), insights into extraterrestrial biology and society, or simply the confirmation that we are not alone in the universe. Of course, this knowledge could also be gained by waiting to receive an interstellar transmission without a substantial investment in transmissions. The problem, of course, is that if everyone in the galaxy is listening and no one is broadcasting then contact will never occur at all. Even if contact is never established, however, investment in METI can still be justified as an archival project whose aim is to preserve a record of human culture by transmitting it into the cosmos should human civilization face extinction as a result of climate change, nuclear war, asteroid impact, or some yet unknown cause.

Given that past transmissions are unlikely to be received, much less understood, by extraterrestrials, the argument that these interstellar broadcasts were a waste of resources is justified only in a narrow sense. Arguably, their value in terms of public engagement and the advances they prompted in message design made them worth it, even if the energy expenditure for the transmissions won't "pay off" through first contact. The point here is that METI cannot be reduced to broadcasting activities. Probably the most important aspect of METI is the design of the message content, a research program that has made great strides in the past few decades with only modest investments. Research on message design is also closely aligned with the aims of SETI insofar as it creates a framework for interpreting an extraterrestrial signal should one ever be received. Furthermore, it helps establish protocols and standards for drafting a response to a received transmission, which could be one of the most contentious events in human history. Laying the theoretical groundwork in advance of this event may prove invaluable in the future. This isn't to say that transmissions can't be designed to be more cost-effective, however. Many cost-benefit analyses of METI fail to consider the rapid adoption of renewable energy technologies such as wind and solar. Other technological advances, such as commercial nuclear fission reactors or so-called room temperature superconductivity would result in the rapid decrease in energy costs and increases in sensor sensitivity to the point where powerful, continuous interstellar broadcasts would represent a trivial expenditure. In the meantime, one possible solution is to forgo informational messages entirely and focus on beaconing. A dedicated optical or microwave beacon, for instance, could be used to target thousands of individual stars for fractions of a second each to draw attention to Earth in order to facilitate communication (Benford, Benford, and Benford 2010a; Ridpath 1978).

WHO SPEAKS FOR EARTH?

One of the most difficult questions for METI was originally posed by Carl Sagan: "Who speaks for Earth?" Interstellar communication is not a conversation between individuals, but entire civilizations, which may exist on planetary or galactic scales. Thus, critics of METI argue that any message sent from Earth must be crafted with the input of all its members. This critique has haunted interstellar communication ever since the plaques affixed to the Pioneer probes were criticized for their ethnocentrism. Although various METI projects have attempted to rectify this shortcoming by crafting more representative messages, such as the Voyager records or the Cosmic Calls and other crowd-sourced messages, the decision to broadcast these messages, as well as their fundamental design, was made by only a handful of individuals. Rectifying this shortcoming is no small task. This text has mostly focused on efforts to eliminate species bias (homocentrism) in interstellar messages by tracing the foundations and boundaries of epistemic universals. As we saw, in many cases even the most ostensibly "objective" knowledge, such as mathematics, can be attributed to the idiosyncrasies of embodied human cognition. A related and equally challenging issue is eliminating cultural bias (ethnocentrism) in interstellar messages. Numerous solutions to this problem have been proposed, but few are satisfactory. One possible solution is to pull all the data from the world's server farms and send the entire internet. Other solutions propose message designs based on aspects of human existence that are present in all cultures, such as music or altruism (Vakoch 2001, 2002, 2010).

A full discussion of message content is beyond the scope of this book, but considering this issue in an abstract sense raises a profound and unsettling question that anyone designing an interstellar message must grapple with: do we want to tell extraterrestrials the truth? Given that even the most well-intentioned message will give only an approximate representation of the diversity of life on Earth, it may be desirable to explicitly address this shortcoming in the message itself. This would mean acknowledging not only our capacity for altruism, but also the fractured nature of our cross-cultural and interpersonal relationships. It would reveal that despite our towering artistic, political, and scientific achievements, global wars and massive economic inequality are fueled by dogmatic belief systems, bigotry, ignorance, and greed. So far, our messages into space have been filtered through rose-tinted glasses. If extraterrestrials saw the whole picture, would they still have any desire to talk to us?

As far as we can tell at this point in history, Earth exists in a galactic backwater. Our stellar neighbors are few and far between and if there are intelligent inhabitants in neighboring solar systems, their silence is conspicuous. Our physical isolation need not condemn us to loneliness, however. Even if a global consensus establishes a moratorium on transmissions, the art and science of designing interstellar messages will never cease to be valuable for what it can teach us about ourselves. Each message, regardless of whether it is broadcast, is like a mirror that reflects the spirit of the age that crafted it. Insofar as the design of interstellar messages demands cross-cultural dialogue and a desire to understand the perspective of someone who may be radically different, it is a research paradigm that may also turn out to be a vector for ameliorating the very terrestrial tensions and divisions that we are so reluctant to reveal to extraterrestrials. The content and structure of future interstellar transmissions will likely be as diverse as the species that created them, but they will all be united by a single vision:

Per linguam ad astra.

APPENDIX A: THE ARECIBO MESSAGE

On November 16, 1974, the first message intended for extraterrestrial recipients was broadcast from the Arecibo radio telescope in Puerto Rico toward Messier 13, a collection of hundreds of thousands of stars located about 25,000 light years from Earth. The message was broadcast as part of a ceremony celebrating a major upgrade to the telescope and the target was chosen because it happened to be visible in the sky during the ceremony. Given that it would take approximately 50,000 years from the time the message was sent to receive a reply, the Arecibo message was more of a technological demonstration of the telescope's new capabilities than an earnest attempt to establish contact with an extraterrestrial civilization. The message was sent at 2,380 MHz and its contents were encoded by modulating the frequency by 10 Hz. The broadcast used 450 kW of power and was transmitted at 10 bits per second so that the entire broadcast lasted just under 3 minutes. The message consisted of 1,679 bits meant to be arranged as a bitmap with 73 rows and 23 columns. The number of bits is a semiprime, a design feature that would inform an ETI how to arrange the message as a picture.

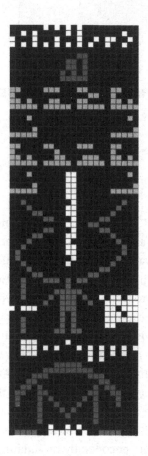

Included in the message were a binary representation of the first ten numbers, the atomic numbers of the five elements that comprise DNA, the formulas for the bases of DNA's nucleotides, the total number of nucleotides according to knowledge at the time, a graphic of DNA's helical structure, a graphic of a human, a graphic of the solar system, and a graphic of the Arecibo telescope, which includes its dimensions. This is a remarkable amount of data given the message's small size (210 bytes), but the encoding scheme is far from intuitive. The binary sequence is encoded left to right and top to bottom. The first four rows consist of binary representations of the first ten numbers ascending from left to right. The fourth row represents the least significant bit such that the first four rows can be read:

ROW	COLUMN																						
	A	B	C	D	E	F	G	H	I	J	K	L	M	N	O	P	Q	R	S	T	U	V	W
1	0	0	0	0	0	0	1	0	1	0	1	0	1	0	0	0	0	0	0	0	0	0	0
2	0	0	1	0	1	0	0	0	0	0	1	0	1	0	0	0	0	0	0	1	0	0	
3	1	0	0	0	1	0	0	0	1	0	0	0	1	0	0	1	0	1	1	0	0	1	0
4	—		—		—		—		—		—		—		—		—		—		—		
	1		2		3		4		5		6		7		8		9			10			

The next portion of the message provides the binary values of the atomic numbers for hydrogen (1), carbon (6), nitrogen (7), oxygen (8), and phosphorus (15). These elements are the basic building blocks of DNA. They are encoded in the message as 4-bit binary numbers with the fifth column reserved to mark the least significant bit. Thus, this portion of the message can be read as:

ROW	COLUMN																						
	A	B	C	D	E	F	G	H	I	J	K	L	M	N	O	P	Q	R	S	T	U	V	W
6	0	0	0	0	0	0	0	0	0	0	0	0	1	1	0	0	0	0	0	0	0	0	0
7	0	0	0	0	0	0	0	0	0	0	1	1	0	1	0	0	0	0	0	0	0	0	0
8	0	0	0	0	0	0	0	0	0	0	1	1	0	1	0	0	0	0	0	0	0	0	0
9	0	0	0	0	0	0	0	0	0	1	0	1	0	1	0	0	0	0	0	0	0	0	0
10	0	0	0	0	0	0	0	0	0	—	—	—	—	—	0	0	0	0	0	0	0	0	0
										H	C	N	O	P									

Next, molecular formulas of nucleotides by providing 3-bit numbers representing the number of each element and the arrangement of these nucleotides on the page is meant to mimic the helical structure of DNA. Thus, for example, the formula for deoxyribose in DNA is C_5H_7O, which would be translated to the bitmap as such:

H	C	N	O	P
1	1	0	0	0
1	0	0	0	0
1	1	0	1	0
—	—	—	—	—
7	5	0	1	0

This represents seven hydrogen atoms, five carbon atoms, zero nitrogen atoms, one oxygen atom and zero phosphorus atoms. In the Arecibo message, these groupings of nucleotide formulas are arranged to show how the nucleotides bond with one another in DNA. So, the nucleotides are arranged in rows 11–30 thus:

DEOXYRIBOSE	—	ADENINE	—	THYMINE	—	DEOXYRIBOSE
PHOSPHATE	—				—	PHOSPHATE
DEOXYRIBOSE	—	CYTOSINE	—	GUANINE	—	DEOXYRIBOSE
PHOSPHATE	—				—	PHOSPHATE

ROW	COLUMN																						
---	A	B	C	D	E	F	G	H	I	J	K	L	M	N	O	P	Q	R	S	T	U	V	W
11	0	0	0	0	0	0	0	0	0	0	0	0	0	0	0	0	0	0	0	0	0	0	0
12	1	1	0	0	0	0	1	1	1	0	0	0	1	1	0	0	0	0	1	1	0	0	0
13	1	0	0	0	0	0	0	0	0	0	0	0	0	0	1	1	0	0	1	0	0	0	0
14	1	1	0	1	0	0	0	1	1	0	0	0	1	1	0	0	0	0	1	1	0	1	0
15	1	1	1	1	1	0	1	1	1	1	1	0	1	1	1	1	1	0	1	1	1	1	1
16	0	0	0	0	0	0	0	0	0	0	0	0	0	0	0	0	0	0	0	0	0	0	0
17	0	0	1	0	0	0	0	0	0	0	0	0	0	0	0	0	0	0	0	0	1	0	0
18	0	0	0	0	0	0	0	0	0	0	0	0	0	0	0	0	0	0	0	0	0	0	0
19	0	0	0	0	1	0	0	0	0	0	0	0	0	0	0	0	0	0	0	0	0	0	1
20	1	1	1	1	1	0	0	0	0	0	0	0	0	0	0	0	0	0	1	1	1	1	1
21	0	0	0	0	0	0	0	0	0	0	0	0	0	0	0	0	0	0	0	0	0	0	0

ROW	COLUMN																						
	A	B	C	D	E	F	G	H	I	J	K	L	M	N	O	P	Q	R	S	T	U	V	W
22	1	1	0	0	0	0	1	1	0	0	0	0	1	1	1	0	0	0	1	1	0	0	0
23	1	0	0	0	0	0	0	0	1	0	0	0	0	0	0	0	0	0	1	0	0	0	0
24	1	1	0	1	0	0	0	0	1	1	0	0	0	1	1	1	0	0	1	1	0	1	0
25	1	1	1	1	1	0	1	1	1	1	1	0	1	1	1	1	1	0	1	1	1	1	1
26	0	0	0	0	0	0	0	0	0	0	0	0	0	0	0	0	0	0	0	0	0	0	0
27	0	0	0	1	0	0	0	0	0	0	0	0	0	0	0	0	0	0	0	0	1	0	0
28	0	0	0	0	0	0	0	0	0	0	0	0	0	0	0	0	0	0	0	0	0	0	0
29	0	0	0	0	1	0	0	0	0	0	0	0	0	0	0	0	0	0	0	0	0	0	1
30	1	1	1	1	1	0	0	0	0	0	0	0	0	0	0	0	0	0	1	1	1	1	1

Directly below the formulas for the nucleotides is a graphic of a helix. In the center of the helix are two columns representing a string of binary numbers (1111111111110111111111101101 011110) which corresponds to the number 4,294,441,822, the number of base pairs estimated to comprise the human genome in 1974. (The Human Genome Project has since revealed that the number of base pairs is closer to 3.5 billion.) Below the helix is a depiction of a human body. To the left of that graphic are four vertical bits above three vertical bits with a horizontal 5-bit sequence in the middle. The vertical bits are intended to represent a yardstick and the horizontal bits represent the binary value 1110 with the leftmost bit indicating the least significant bit. This binary value corresponds to the number 14, which can be multiplied by the wavelength of the message (126 millimeters) to get the average height of a human male: about 1.76 meters. That 14 needs to be multiplied by the wavelength of the message is not explicitly demonstrated in the message, so it is unlikely that extraterrestrials would be able to decipher this portion. To the right of the human body is a binary number (0000111111 11110111111011111111110110) that represents 4, 292,853,750 in decimal format and corresponds to the approximate population of the Earth in 1974. This block of bits is read from the bottom right corner moving right to left. The least significant bit is indicated by an offset bit in the upper left-hand corner of the block.

Below the human is a schematic depicting our solar system. The Sun is depicted on the left as a 3×3 array of bits. Mercury, Venus, Earth, and Mars are each represented with a single bit, with the Earth's bit shifted up to indicate that we reside on the third planet. Jupiter and Saturn are represented by three vertical bits, Uranus and Neptune by two vertical bits, and Pluto by a single bit. These different bit-values are meant to give an approximate scale of the various planets and the Sun.

Finally, there is a graphical representation of the Arecibo telescope. The bottom two lines of the message are meant to give the dimensions of the telescope. Two horizontal lines of four bits each serve as a ruler whose value is indicated by the binary number in the middle. The binary number consists of two rows and is read left to right from the upper left corner to the bottom right corner. This returns a value of 100101111110 that corresponds to 2,430 in decimal format. When this number is multiplied by the wavelength of the transmission, it returns 306.18 meters, the approximate diameter of the Arecibo telescope.

APPENDIX B: THE COSMIC CALL TRANSMISSIONS

The first Cosmic Call message was sent from the Evpatoria radar in Ukraine on May 24, 1999. The message was sponsored by Team Encounter, a Texas-based company that crowd funded the transmission by allowing the public to buy space in the message for short personal messages. The message had a two-tiered structure: The first tier consisted of four scientific messages, one of which was designed by the Canadian physicists Sté-phane Dumas and Yvan Dutil, another designed Team Encounter employee Richard Braastad, a staff message from employees at Team Encounter, and the original Arecibo message. Each of these scientific messages was broadcast three times at 100 bits per second and the public portion of the message was broadcast once at 2,000 bits per second. The transmission was broadcast at 5.01 GHz and encoded by modulating the frequency by 24 kHz. The power used for the broadcast was 150 kW. The message targeted five stars within seventy light years of Earth. Here we will consider only the design of the Dumas–Dutil portion of the message.

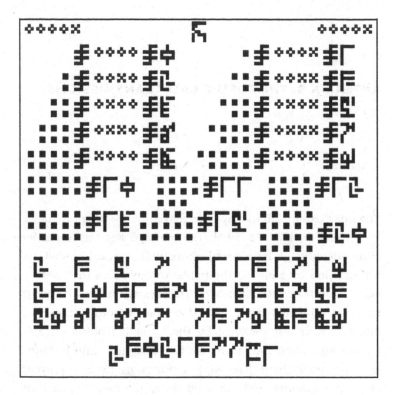

The Dumas–Dutil message consists of twenty-three "pages," each of which is arranged as a 127×127 pixel array. Each page is delineated by a one-pixel border and the page number is given in binary in the upper left- and right-hand corner of the page. Thus, on page 1 shown in the figure, there are five symbols in each upper corner corresponding to 00001. The large-scale message structure was inspired by Freudenthal's *lingua cosmica* and begins with a basic introduction to arithmetic before proceeding to discuss more complex topics such as the elements, units of measurement, our solar system, and so on (a complete list of the topics of each page is provided below). Each symbol consists of a 5×7 bitmap designed to be resistant to error. In other words, changing the bit-value in one symbol shouldn't turn it into a different symbol. A complete dictionary of the symbols used in the 1999 Cosmic Call message is provided below. The first page of the message introduces numbers and the concept of identity. The following equation, for example, portrays the concept of "four":

On the left hand side of the equation is an ostensive representation of four featuring four squares. The symbol ⧧ corresponds to "=", the crosses in the middle represent the binary notation for four (0100) and the rightmost symbol ⊑ corresponds to four. Thus, the equation reads "4=4=4," with the number four provided in three different formats. Importantly, the system uses a base-ten positional notation system. Thus "10" would be written by combining the symbol for 1 (⌐) with the symbol for 0 (⧫) to produce ⌐⧫. At the bottom of the first page are the first twenty-four prime numbers and the largest prime number known in 1999, stylized as "$2^{3,021,377}-1$."

MESSAGE CONTENTS

Page 1: Numbers
This page defines the numbers 0 through 10, as well as 11, 12, 14, 15 and 20 using ostensive definitions, binary code, and message-specific symbols. The first twenty-four primes are introduced, as well as the largest prime number known in 1999.

Page 2: Arithmetic Operations
This page introduces the symbols for addition, subtraction, multiplication, and division operators by using simple equations such as "1 + 1 =2." It also introduces fractions and negative numbers.

Page 3: Exponents
This page introduces the notation for exponents by using number symbols as superscripts.

Page 4: Variables
This page introduces the concept of variables by posing questions. It also features a graph with labels for the X and Y axes.

Page 5: Geometry
This page introduces the concept of pi by showing a graphical representation of a circle and its radius. The first seven digits of pi and the last fifteen digits known in 1999 are given. It also includes a graphical representation of the Pythagorean theorem.

Page 6: Elements
This page defines symbols for proton, electron, and neutron. It shows a graphical Bohr model of the hydrogen atom and gives the mass of its proton and electron. The mass of the proton is defined in relation to the mass of the electron. It defines ten elements (hydrogen, helium, carbon, nitrogen, oxygen, aluminum, silicon, iron, sodium, and chlorine) by enumerating the number of protons and electrons for each element.

Page 7: Mass

This page defines the concept of a kilogram by defining the weights of protons, neutrons, and electrons in this unit of measure. Avogadro's number, the number of atoms in a mole, is defined. Six other elements are also introduced (sulfur, zinc, argon, uranium, silver, and gold) by enumerating their protons and electrons. Element 112, copernicium, is also defined to demonstrate how much of the periodic table had been discovered so far. (The existence of element 114, flerovium, was announced in 1998, but hadn't been confirmed by the time the message was sent.)

Page 8: Hydrogen Atom

This page discusses the spectrum of the hydrogen atom, which is used to introduce symbols for frequency, wavelength, time, and energy. The speed of light is also defined by giving the relationship between its frequency and wavelength.

Page 9: Units of Measurement

This page defines force, energy, pressure, and power, as well as their respective units of measurement (Newton, Joule, Pascal, and Watt). It also defines speed, acceleration, Planck's constant, and the gravitational constant.

Page 10: Temperature

This page gives the boiling and melting temperatures of six elements (hydrogen, carbon, sulfur, zinc, silver, and gold) in Kelvins. It also includes a graph showing the boiling and freezing points of water.

Page 11: The Solar System

This page gives a graphical representation of our solar system and attributes a symbol to each planet. The mass and radii of Jupiter and the Sun are also given.

Page 12: The Earth and Moon

This page gives the mass and radius of the Earth and Moon and the distance between them with the help of a graphical representation. It also gives the distance between the Earth and the Sun.

Page 13: The Earth and Moon

This page defines the length of a day and year on Earth, the orbital period of the Moon, and the ages of the Earth and Sun.

Page 14: Earth's Crust, Atmosphere, Water, and Gravity

This page features a graphical representation of three people standing on land next to a mountain that descends into the ocean. It defines the main elements found in Earth's crust, atmosphere, and ocean. It also gives the height and depth of the highest and lowest points on Earth, as well as the acceleration due to gravity at Earth's surface.

Page 15: Human Appearance

This page depicts a nude male and female human. The graphic is similar to the drawing of humans included on the Pioneer plaque and the male human is seen raising his arm in a gesture of greeting. It defines the height of these humans as 1.8 meters.

Page 16: Human Physiology

This page defines the range of human visual and auditory perception, our average body temperature, average mass, average lifespan, and the number of people on Earth at the time (approximately 6 billion).

Page 17: DNA
This page uses a graphical representation of DNA to introduce the four nucleotides: thymine, adenine, guanine, and cytosine.

Page 18: Cells
This page shows a graphical representation of a cell. A helical structure is shown in the nucleus of the cell. Further, it defines the approximate number of cells in a human (10^{13} cells) as well as the average cell size.

Page 19: Map of Earth, Left Fuller Projection
This page shows a graphical representation of the Earth as a Fuller projection. Oceans and landmasses are differentiated with symbols introduced on page 14. This page only features the left side of the Fuller projection and depicts Australia, Asia, Europe, and Africa.

Page 20: Map of Earth, Right Fuller Projection
This page provides the other half of the Fuller projection introduced on page 19. It depicts North America, South America, and Antarctica.

Page 21: Evpatoria Radar Parameters
This page describes the capabilities of the Evpatoria radar used to send the Cosmic Call message. It defines the frequency of the message, its wavelength, the number of people who contributed messages to the transmission (43,000), the power of the transmitter (150 kW), and its radius (70 m). It also introduces

a symbol for "you," implying the extraterrestrial receiver. This symbol is used on page 23 to ask the recipient questions.

Page 22: Cosmology

This page describes basic cosmological knowledge such as the expansion of the universe, the density of the universe, the cosmological constant, Hubble's constant, and the temperature of the universe.

Page 23: Questions

This page lists questions for the recipient. These include questions about their mass, their planet, their biology, their units of measurement, and so on.

COSMIC CALL DICTIONARY

Each of the symbols used in the 1999 Cosmic Call message consisted of a 5×7 bitmap. The symbol is given on the left and its meaning on the right.

COSMIC CALL 2003

The second Cosmic Call was broadcast to six nearby (< 50 light years) Sun-like stars on July 6, 2003, from the Evpatoria radar in Ukraine. Like its predecessor in 1999, the second Cosmic Call had a two-tiered message structure. The first tier consisted of five scientific messages and the second tier consisted of thousands of short natural language messages that had been crowd sourced from the public. The five scientific messages included an updated Dutil-Dumas message, the Arecibo message, a "bilingual image glossary," a message from the Team Encounter staff,

▨	1	▨	2	▨	3	▨	4
▨	5	▨	6	▨	7	▨	8
▨	9	▨	0	▨	Newton	▨	Joule
▨	Addition	▨	Equal	▨	Math	▨	Pascal
▨	Subtraction	▨	π	▨	Radius	▨	Watt
▨	Multiplication	▨	Union	▨	Kilogram	▨	Hertz
▨	Division	▨	Decimal point	▨	Meter	▨	Kelvin
▨	Undetermined	▨	Delta	▨	Second	▨	Year
▨	Negation	▨	Hydrogen	▨	Helium	▨	Carbon
▨	Nitrogen	▨	Oxygen	▨	Aluminum	▨	Silicon
▨	Iron	▨	Sodium	▨	Chlorine	▨	Argon
▨	Element 112	▨	Gold	▨	Silver	▨	Sulfur
▨	Uranium	▨	Zinc	▨	Proton	▨	Neutron

	Electron		Mass		Wavelength		Time
	Frequency		Velocity		Force		Energy
	Pressure		Power		Planck's constant		Gravitational constant
	Distance		Hubble constant		Density		Cosmological constant
	Acceleration		Charge		Length		Physics
	Temperature		Photon		Thymine		Guanine
	Adenosine		Cytosine		Guanine		Cell
	Biology		Male		Female		People
	Jupiter		Earth		Moon		Sun
	Mars		Mercury		Neptune		Pluto
	Saturn		Uranus		Venus		Etc. / ...
	Universe / Cosmology		Variable / Question		Land		Ocean
	Atmosphere		"You" / Recipient		Age		Var a
	Var b		Var c				

and a message designed by Team Encounter's message lead Richard Braastad that used the symbolic system created by Dutil and Dumas to describe a spaceship that Team Encounter planned to build called *Humanity's First Starship*.

Unlike the first Cosmic Call message, which played each scientific message once and then repeated that cycle three times, Cosmic Call 2003 did three broadcasts of the Dutil-Dumas message, Arecibo message, and the bilingual image glossary back-to-back before moving to the next message. The Braastad message, Team Encounter message, and the public portion of the message were only broadcast once. The message was broadcast at 5.01 GHz with a power of 150 kW. The scientific portion of the message was broadcast at 400 bits per second and the public portion of the message was broadcast at 100 kilobits per second. Here we will consider the updates made to the Dutil-Dumas message and the structure of the bilingual image glossary.

Unlike the Dutil-Dumas portion of the 1999 Cosmic Call, in 2003 their message was formatted as one long page, but a one-pixel border was still included. This new format was not only more space efficient, but the presence of two single pixel vertical lines running the length of the message as a border also helped ensure that that the message would be decoded correctly. The page was divided into sections marked with binary numbers and each section retained most of the content of the 1999 message. Perhaps the most significant change was the symbols for numbers 0–9 were reduced from 5×7 bitmaps to 4×7 bitmaps. This helped save space in the message owing to the frequency that these symbols were used throughout the message and helped distinguish numerals from other symbols.

The bilingual image glossary included in the 2003 Cosmic Call transmissions was originally designed as part of the Teen

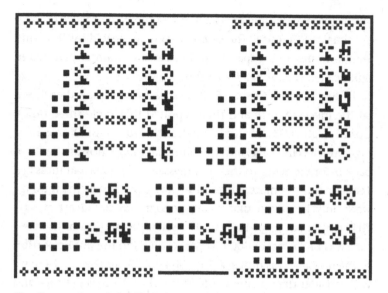

The first section of the 2003 Cosmic Call message describing numbers. The numbered section dividers can be seen at the top and bottom of the image.

Age Message transmission sent from the Evpatoria radio telescope in 2001. For the Teen Age Message, the bilingual image glossary consisted of twelve binary images rendered as bitmaps of 101×101 pixels. The images depicted the following concepts: "children," "parents," "man," "people," "family," "galaxy," "Earth," "nature," "Sun," "house," "game," and "send message," a request for a reply, and a depiction of the Evpatoria radar. Each image included a description of the picture written in English and Russian. For the Cosmic Calls, all the images were combined to form a single "page," and their horizontal length was shortened by one row except for the final image. Thus, the entire image glossary consisted of 1,201 lines of pixels. The total size of the file was 121,301 bits, the product of the primes 101 and 1201.

Finally, it is worth mentioning the contents of the public portion of the message. These included thousands of messages that were digitized and put into twenty-four folders according to topic. Among the topics included in the 2003 Cosmic Call message were: the entire contents of the HelloToETI website, which was created by the futurist Allen Tough; pictures of flags from 282 countries and organizations; the song "Starman" by David Bowie; an album and photos from the Hungarian rock band KFT; pictures drawn by Ukrainian children; and the text of a bill adopted in New Mexico that designated the second Tuesday in February as "Extraterrestrial Culture Day." This portion of the message was transmitted at 100,000 bits per second. This is far faster than the 400 bits per second used for the scientific portion of the message, but this was of necessity. Had the public portion of the message been transmitted at the same rate as the scientific message, the transmission would've taken over fifty days of continuous broadcasting. Even at this greatly increased bitrate, however, the public portion of the message still took approximately eleven hours to transmit, which makes it the longest continuous broadcast of any message into space thus far.

APPENDIX C: LINCOS

Its name a portmanteau of *lingua cosmica*, *Lincos* was written by the Dutch mathematician Hans Freudenthal as a first attempt at designing a language specifically intended to foster communication between Earth and extraterrestrials. There are several considerations to take into account when appraising *Lincos*, which is composed as a compendium of "texts" that are meant to be sent as packages to an extraterrestrial receiver. In the first place, *Lincos* is not an actual program for transmission. Rather, it is something of a roadmap for designing such a transmission, offering a syntax and lexicon that can be drawn upon should an actual message be modeled on the Lincos system. As such, *Lincos* is far shorter than an actual program would be, insofar as it only provides enough examples for each novel scenario to make the didactic approach to conveying those messages clear to the reader. A real program would have several more examples for each instance since the receiver cannot be presumed to find these transmissions as intuitive as a human would. Moreover, the chapters found in *Lincos* are grouped to make sense to the human reader—that is, by categories—although these wouldn't necessarily be the ideal groups for composing a message to extraterrestrial beings.

Second, we need to distinguish between Lincos proper and the examples of Lincos given in Freudenthal's book, a distinction that can be conceived as analogous to the difference between a language when it is spoken and written. Lincos was designed as a spoken language, which means that it traffics in phonemes rather than letters, and phonetics rather than spelling. The phonemes of Lincos are differentiated as radio waves of varying length and duration, although how these phonemes would be coded as radio waves was not explicitly addressed by Freudenthal, and their written coded version (as seen in this book) is comprised of arbitrary symbols that bear no necessary relationship to the message being sent. The symbols used in the coded Lincos here is a goulash of Arabic numerals, scientific symbols, and Latin abbreviations. Thus, as Freudenthal explains, "the spoken Lincos word corresponding to 'Num' need not be composed of phonetic groups which correspond to N, u, and m. ... We shall stick to the convention that, e.g. 'Num' is not the proper Lincos word but the written (or printed) image of it." That said, it is worth noting that the phonetic system of Lincos is systematic so as to make it more easily understood by terrestrial readers.

Freudenthal argues that the difficulty in communicating with an intelligent being who shares none of our natural languages is not quite as foreign as the reader might imagine it to be. In fact, he argues, this is done every day with our communication with infants, who are a "tabula rasa" of lexicographic and syntactic knowledge. Admittedly, this approximation is not as great as it at first seems. In the first place, modern linguistics suggests human infants are not a tabula rasa, but are born with an innate language faculty, which makes them predisposed to learn their native language(s) over time. Moreover, Freudenthal acknowledges that a lot of early linguistic development occurs by showing—pointing to a tree and saying "tree," a cat and saying "gato," and so on. Obviously, this tactic of instruction through showing will not be available in the astrolinguistic context. Even sending visual aids—whether still images of television programs—will first require the transmission of instructions showing how to "unpack" the signal so that it can be properly seen. As Freudenthal rightly points out, by the time such communicated instructions would be intelligible to the alien recipient, these visual aids would be mostly superfluous.

So then, where to begin? For Freudenthal, the ideal starting place was with numerals. Just as the earliest examples of human writing were ideogrammatic (where a symbol refers to the thing it depicts), so too will the first Lincos texts be "ideophonetic" insofar as they will consist of short peeps where the number of peeps corresponds to the numeral being represented. In this sense, Freudenthal is violating his own dictum to avoid reliance on "showing" rather telling, but in his *lingua cosmica* such a shortcoming can hardly be avoided. As such, he is depending on the intelligence of the recipients to initially recognize that n peeps correspond to the numeral n. After learning to count to ten, we may begin to teach basic mathematical operators, such as addition and subtraction, or the principle of identity. In Freudenthal's shorthand, this would look like this:

.... > ...

..... = ...

Next, these ostensive numerals (which function as a "picture" of a number they represent) are superseded by binary equivalents.

. = 1 .. = 10 ... = 11 .... = 100

The program would then repeat the same examples used to introduce operators with ostensive numerals with binary numbers to
establish their equivalence. So far, Freudenthal's *lingua* bears the
hallmarks of previous attempts at designing an interstellar language such as the Astraglossa or Galton's telegraphic system. Its
first major departure is with the introduction of variables. Consider the following example:

$$100 > 10100 + 11 > 10 + 11100 + 1101 > 10 + 1101100 + 1$$
$$> 10 + 1100 + 110 > 10 + 110100 + 11111 > 10 + 11111100$$
$$+ a > 10 + a$$

By using examples such as these, and incorporating multiple
variables into successively elaborate texts, Lincos will convey the
notion of a mathematical variable to the extraterrestrial recipient. Variables will in turn be used to introduce logical connectives, such as

$$a + b = c \rightarrow c - b$$

Importantly, Freudenthal notes that he hasn't introduced binary
numbers into examples showing logical connectives, since the
above example, if rendered as $10 + 11 = 101 \rightarrow 10 = 101-11$,
would be meaningless if the recipient didn't understand formal
implication. Freudenthal then goes on to introduce the concept
of tautologies as well as a symbol denoting a question, which he
laments is absent from formal systems of symbolic logic based
on "the philosophy that mathematics is rather a stock of true
propositions than an art of discovering." In his shorthand, a
question soliciting the value of x would look like this:

$$?xx + 101 = 11$$

After introducing multiplication and division, Freudenthal
introduces a symbol for the set of all natural numbers ('Num'), a

symbol for the set of integers ('Int'), as well as a symbol indicating set membership ∈. These are introduced through repetition:

1 ∈ Num

10 ∈ Num

111 ∈ Num

And so on. A symbol for the universal quantifier (denoting "for all") is also introduced (Freudenthal lists it as ∧, although today ∀ is the more conventional symbol for this relation and we will substitute this standard symbol for his), as well as an existential quantifier ("there exists"), for which we shall also use conventional notation, ∃. Using these tools, Freudenthal proceeds to define prime numbers, rational numbers, real numbers, and so on. He also introduces a name for "sets" (Agg), which is introduced by way of examples since the notion of a set cannot be formally defined within the system without a highly developed metalanguage for discussing and formally defining elements of that system. Interestingly, Freudenthal introduces a symbol designating a definite article in the mathematics section, even though he acknowledges that "the factual use of the article can only be learned in non-mathematic contexts." From this it also follows that "we cannot decide in a mechanical way or on purely syntactic grounds whether certain expressions are meaningful or not. But this is no disadvantage. *Lincos has been designed for the purpose of being used by people who know what they say*, and who endeavor to utter meaningful speech" (emphasis mine). With set theory now introduced, Freudenthal introduces the idea of a function as a special instance of a set, but immediately notes this is inadequate. As he rightly points out, functions are often thought of as "kinematic laws" that map the value of a variable x to the value of a variable y. Freudenthal fears that his introduction of a function at this point would be premature, since its introduction as a special set would not do justice to importance of the notion of a function, and at this point Lincos lacks the language to drive home that point. He suggests it may be better to introduce the notion of a function in the "Behavior" section of Lincos, perhaps after introducing the concept of "law." Finally, the mathematics section is rounded out with the introduction of a few words that will be used in more complicated expressions about life on Earth, namely Ver (true), Fal (false), Prp (proposition), Qus (questions), Iud (truth value), and the symbol for negation, ¬. Conspicuously absent from the mathematics section is a description of calculus, which Freudenthal acknowledged as a crucial element of an interstellar message in practice, but he opted not to pursue its notation in *Lincos*. This, he writes, is because "it is a fact that the usual system of notations of Calculus is inconsistent to a degree, especially in the field of functions of two and more variables. It is easy to remove these defects, but a notational system which preserves the obvious advantages of Leibniz's notation is still lacking." One can hardly blame Freudenthal for not wanting to reinvent the notation of calculus or burden his readers with a textbook of calculus written in Lincos notation.

Following Freudenthal's description of mathematics, we move into their concrete application to notions of time, behavior, and science. I will address these only very briefly, on the grounds that they are merely applications of symbolic logic set forth in the notation described in the chapter on mathematics. A few examples should suffice to give the reader a taste of how Freudenthal envisioned the construction of a message for extraterrestrials that used Lincos notation. For instance, Freudenthal introduces the notion of time using the radio signals themselves, which are communicated in his shorthand as dashes of varying lengths. Thus:

Dur———= Sec a

indicates a radio signal of a duration———, which is equivalent to "a" seconds. In this example, "a" is a metavariable indicating the duration of the signal that is separate from the variables introduced in the section on mathematics. In terms of syntax, "Dur" is treated as a function belonging to the set of the duration of radio signals in the Lincos message. "Sec," on the other hand, works a function mapping the set of positive integers (Pos) to the set of durations to establish the notion of the time unit we call a second. This is also how other units of measurement, for instance a centimeter (Cmt) or a gram (Grm), will be introduced, namely as functions mapping the set Pos to the set of measurements. Using this as a basis, Freudenthal introduces notions of frequency and its unit of measurement (Hertz), as well as the notions of "before," "after," "beginning," and "end."

In the third chapter, Freudenthal dives into the messy business of communicating facts about human behavior using the notations roughly sketched out above. This is done by way of "morality plays," in which Freudenthal posits the existence of two or more humans who discuss some event that corresponds to whatever behavior is under consideration. Thus, Lincos can communicate ideas about very abstract topics such as love, death, and individuality. Importantly, Freudenthal is *showing* the extraterrestrial recipient examples of aspects of human behavior, rather than trying to define general examples in the systematically rigorous way in which he introduced mathematics and time-related concepts. He anticipates that, based on these individual examples depicting human behavior in one event, the extraterrestrial will be able to generalize further about human behavior even though general "rules" of human behavior have not been provided. This is a defining characteristic of Freudenthal's *lingua* and indicative of his approach to teaching more generally.

To introduce behavior, Freudenthal adds dozens of new words to his vocabulary: Inq (indicating someone is speaking), Hom (the set of humans), Tan (nevertheless), Qqm (though), names for individual humans (stylized as Ha, Hb, Hc and so on), and many others. As Freudenthal notes, in Lincos, people do not have consistent personalities (so that Ha appears multiple times in the dialogues and acts quite contrary to his previous actions on many instances), something that Freudenthal argues would need to be addressed in an actual transmission so that each instance of a character in the plays has a consistent personality. As far as the dialogues themselves, they take a basic form of

Ha Inq Hb σ

which can be read as "Human a tells Human b some thing σ." To introduce this format, Freudenthal's characters discuss mathematical objects that have already been defined in previous examples. Up to this point, mathematics and time did not require truth-values, since only true examples were used. Thus, rather than sending a message that claims 1=2 is false, Freudenthal added a symbol to his vocabulary (≠), so that 1≠2 maintains the use of only true statements for examples. Human behavior, however, is not so simple and cannot be communicated only with reference to "good" actions. To place values on actions, he cannot use "Ver" and "Fal" since those attribute truth-values to propositions. To this end he introduces "Ben" and "Mal," which are the valuations "good" and "bad," respectively.

Since Freudenthal has not set out to write a textbook on morality, these valuations of actions are not introduced as behavioral *rules*, but rather judgments issued by the characters in the morality plays. For example, here is a dialogue in which Human a asks Human b about the identity of some person *x* (Human c) who is present, and is then introduced to Human a by Human b. The lamba is used to indicate a definite article (in this case, "the *x*"). T_1 and T_2 are used to indicate a duration of time during which the event described between the two markers occurs. So:

Ha Inq Hb ? *x*. *x*= Hc	Ha asks Hb, who is *x*? (*x* is Hc)
T_1 Hb Inq Hc Hc T_2	Hb speaks to Hc and says "Hc"
Hb Inq Ha: λ*x*. T_1 T_2 Hb Inq *x* Hc	Hb tells Ha: The *x* I spoke to between T_1 T_2 is Hc
Ha Inq Hb Ben	Ha tells Hb "good"

The translations on the right-hand side are approximate, but the reader should get a good idea of what has occurred—namely, Human b has introduced Human a to Human c after Human a inquired who that person "*x*" is.

In the *Lincos* text, Freudenthal provides hundreds of examples like the ones above, methodically developing his outline of a language intended for extraterrestrial recipients. Yet for all its robustness, he was the first to admit the areas it was lacking, such as the Lincos lexicon. For Freudenthal, these oversights or shortcomings were an unavoidable symptom of trying to navigate the imprecision of natural languages and the total inflexibility of formal "logistic" languages in Lincos.

Although Freudenthal admired the ability of natural languages to communicate meaningfully without relying on rigid formalisms, his language still more closely resembles formal, logical languages than their natural counterparts. However, the singular hybridity of Lincos didn't have anything to do with Freduenthal's belief in any particular philosophy of language. "Attempts to construct symbolic languages have often been linked to some philosophy, though the linking usually proved to be less close afterwards than the author believed," Freudenthal wrote in the 1957 introduction to *Lincos*. "Dealing with language as a means of communication has been the only philosophy I have espoused in the present project."

APPENDIX D: THE LAMBDA CALCULUS AND ITS APPLICATION TO ASTROLINGUISTICS

If Freudenthal's original Lincos corresponded with the mathematical turn in the philosophy of language in the early twentieth century, then Alexander Ollongren's new system, which he also called Lincos, was a product of the digital computing revolution. Developments in mathematical logic, particularly the lambda calculus and the early programming languages it begat, made it clear that a more efficient linguistic system for interstellar communication was possible. Thus, rather than designing a semiformalized language based on arithmetic, as Freudenthal had done, Ollongren chose higher-order logic as the basis of his Lincos.

For Ollongren, logic is an ideal starting point for an interstellar language because it seems reasonable to assume that any extraterrestrial society capable of sending and/or receiving interstellar messages must, to some degree, be familiar with logic. But logic, which following Ollongren (2013) we may define as the "the correct reasoning over abstractions of reality," is not some monolithic entity—there are many types of logic, each with its own strengths and weaknesses. So, as Ollongren was aware, the problem of *which* logic to use as the basis for a cosmic language presented itself as an immediate problem.

Two of the better-known modalities of logic are the propositional and predicate calculi. Propositional logic has a relatively
simple syntax in which expressions are created from propositions joined by four operators: and, or, implication, and negation. As we saw previously, the meaning of these propositions is
truth-functional and laid out in the well-known truth-tables. Yet
as Ollongren notes, this is not a suitable basis for an interstellar
language since it is not able to make assertions about abstractions over classes of objects that are united by some predicate,
only assertions about abstractions over individual objects.

In this sense, predicate logic is a stronger calculus insofar as it
incorporates quantifiers such as "all" and "exists," which allow
for abstractions over classes of objects satisfying some predicate.
The trade-off, of course, is that the syntax and semantics of the
expressions becomes more complex. When designing a linguistic system for interstellar communication, however, one of the
goals of the language should be to remain as syntactically simple
as possible, while still retaining robust semantic potential.

For this reason, Ollongren turned to constructive, or intuitionistic, logic and its underlying theory of types. Constructive
logic significantly departs from classical logic by its rejection
of the law of excluded middle, which states that a proposition must be either true or false. Historically, this rejection
of this main tenet of classical logic has been the source of
some controversy—the mathematician David Hilbert famously
described this limitation as akin to asking an astronomer to
work without a telescope (Hilbert 1928). Nevertheless, Ollongren demonstrated that constructive logic has immense practical use when it comes to designing a language for interstellar
communication.

LAMBDA CALCULUS AND CONSTRUCTIVE MATHEMATICS

The calculus of constructions with induction at the core of Ollongren's Lincos is the result of wedding two formal systems that for a large part of their developmental history were entirely separate from one another—constructive mathematics and the lambda calculus. Both were developed in the course of investigations into the foundation of mathematics, but the lambda calculus quickly became bound up in the computing revolution, while constructive mathematics developed independently as a foundational theory of mathematics.

The lambda calculus (henceforth referred to as λ-calculus) is a logical system developed by Alonzo Church in the early 1930s as a method for formalizing mathematical logic. It has its roots in Bertrand Russell's work on type theory, although the λ-calculus was originally untyped, which means that relevant sets for a function are not made explicit (see Church 1932, 1936). Before delving into the differences of typed and untyped λ-calculus, let us consider the formal elements of the system. Expressions in the lambda calculus either consist of variables (x, y, z ...), constants (such as integers or Booleans) or fixed functions requiring at least one argument (variables can also play the role of a function, albeit one with arity 0). Importantly, there is no variable declaration in the lambda calculus. Rather, the λ is used to bind variables anonymously in a function—that is, to bind the variable without giving the function a name. To clarify the elements of the Lambda grammar in a formal way, Ollongren offers a context-free generative grammar derived from a report on the syntax of the programming language ALGOL-60 (Backus et al. 1963):

Expr ::= constant | variable | function

function ::= expr expr | λ var⁺ .expr

Read informally, the first line indicates that a lambda expression can consist of a constant, variable, or function. The second line indicates that a function consists of an expression applied to another (otherwise known as functional application) or a lambda abstraction (here, var⁺ indicates there can be more than one variable). This formal context-free grammar for the lambda calculus gives a clear picture of its extremely simple syntactical rules.

The purpose of the lambda in a function is to abstract over the variables in an expression, otherwise known as a lambda abstraction. A lambda abstraction is a lambda function that takes a single input (say, y) and applies it to an expression, which is some combination of variables, constants and functions. Formally, a lambda abstraction will be written like this: λy.x, where "y" is the variable and "x" is the expression. For example, λy.y would represent the identity function y→y. A lambda application, on the other hand, represents the application of a function to an input. For example, (λy.y)x would be the identity function applied to the input "x." To make things a little more concrete, consider a simple polynomial expression such as $[(x^2-7) \times x]$.Written as a lambda abstraction, the expression would look like λ x.$[(x^2-7) \times x]$. The lambda has no effect, other than to denote that the "x" here is bound in the expression. Once a value is given for x, say 3, the lambda abstraction may be rewritten as:

$$(\lambda\, x.[(x^2 - 7) \times x])\, 3$$

Informally, this reads that 3 should be applied to the expression in place of x, allowing us to carry out the function: $(3^2-7) \times 3$.

There are two main conversions used in lambda expressions: the alpha-conversion and beta-conversion. For alpha conversions, consider $\lambda y.\ (+y\ 1)$, an expression that represents the notion "add 1." If we write $\lambda x.\ (+x\ 1)$, this represents the same operation, so we could write $\lambda y.\ (+y\ 1) = \lambda x.\ (+x\ 1)$—as is evident from this example, alpha conversions are simply a means of renaming variables in lambda abstractions. A beta-conversion involves substituting bound variables for arguments during an application of a lambda function, just as in the original example above. Now consider a beta-reduction in a lambda expression ranging over two variables:

$\lambda x, y.\ (+x\ y)\ 1\ 2$

Here a lambda abstraction with two variables $\lambda x, y.\ (+x\ y)$ is applied to two arguments (1 and 2). For most of us, the problem is trivially simple: $1 + 2 = 3$. But the advantage of the lambda calculus is that its operation also furnishes a proof and the price to pay for this is currying the expression. So first the value for x would be applied to the original lambda abstraction to result in $(\lambda y.\ (+1\ y))2$. The beta-reduction is then applied once more to arrive at $+\ 1\ 2$ (in prefix notation), which of course is equal to 3.

After some initial turbulence during development of the untyped lambda calculus (Kleene and Rosser 1935), Church developed a simple typed λ-calculus that would only accept data as input that corresponded to the correct type for that function. The typed lambda calculus had the advantage that more functions were able to be *proved* within the system itself. It is this last element that is especially important for its use in the context of interstellar messages, since the receiver must be able to verify the equations in the message without feedback from the sender.

A higher-order typed lambda calculus undergirds Lincos 2.0, so let's consider briefly its formal elements using the original example for alpha conversions, λy. (+y 1). Suppose that we want y to only take arguments over a certain range, such as the natural numbers. This range, which is the type, can then be expressed as [λ y :nat]. (+y 1), indicating that y is of the type "natural numbers." This is a simply typed lambda calculus insofar as only the function is given a type. But in the typed lambda calculus deployed in Lincos 2.0, *all* elements in an expression will be typed. Thus, addition will need a type, configured as +: nat→nat→nat. Informally this reads that the addition symbol represents a type of operator that requires two arguments of type "nat" (i.e., a natural number) that produces a result in type "nat."

This new formalization of mathematical constructivism in the lambda calculus offered two primary advantages over its predecessors. In the first place, the functions were anonymous insofar as they didn't require names, a convenience in its applications to programming languages (for more on this point, see Mitchell 2002). In the second place, each function within the lambda calculus takes only a single input, or argument. Obviously, some functions require multiple inputs. The lambda calculus can handle these functions thanks to a process known as currying, which breaks down functions with multiple inputs into strings of functions that each require a single input.

Since functions within the λ-calculus can be both the inputs and outputs of other functions, this tends to require more steps in computing a function than other computational methods. Although this may be inefficient, its advantages far outweigh the costs insofar as it also lays bare the steps that were used to arrive at the computed answer to a function within the calculus itself. In short, *every function furnishes its own proof.*

Here we see the intersection of the lambda calculus and constructivist logic. As its name suggests, constructive logic approaches reasoning about the abstractions of reality by constructing direct evidence, or proof, for its propositions. Unlike classical logic, where propositions are assigned truth values regardless of proof—that is, their proof is indirect—propositions in constructive logic do not automatically have truth-values assigned. Rather, practitioners must construct direct evidence for a proposition, thereby also furnishing its proof. As Per Martin-Löf further demonstrated, this constructivist method can also be used toward semantic ends, insofar as "the meaning of a proposition is determined by ... what counts as a verification of it" (Martin-Löf 1996).

Such ideas grow out of a critique of the logic employed by the formalists such as Bertrand Russell and David Hilbert, where indirect proofs are furnished about the truth or falsity of propositions involving objects, without showing or "finding" how that object exists in the first place. Rather, the existence of objects invoked in propositions are assumed by first denying their existence and then proving a contradiction on this basis. This proof by contradiction is not allowed in constructivist mathematics, which requires the construction of a proof (of existence) for every mathematical object. In this sense, it may be considered a stronger candidate for the basis of system for interstellar communication. As Freudenthal demonstrated, introducing mathematics to an extraterrestrial intelligence by way of false statements (or in this case, nonexistent objects) opens the door for confusion and error, which is limited by the nature of the constructive system.

Constructivist mathematics is largely dependent on a style of logic known as intuitionism, pioneered by the Dutch mathematician L. E. J. Brouwer. Rather than logic being the foundation of mathematics, Brouwer thought that mathematics was foundational to logic. This was because Brouwer did not conceive of mathematics as something objective, but rather saw it as a subjective activity consisting in the mental activity of individuals. On this view, mathematical objects are created as mental constructions in people's minds based on other mental constructions. In short, people can prove the existence of mathematical objects by constructing them as mental objects. Mathematical language, and its more stringent, context-free form, logical language, was to Brouwer merely the means of describing these mentally constructed mathematical objects to others. But for Brouwer, these logical languages were themselves just series of mathematical constructions. Thus, it was then the task of the logician to articulate the mathematical basis of the logical language (see Brouwer 1981, 1996).

This critique is at the root of intuitionistic logic, which is very much like the classical logic except that it denies the law of excluded middle, which states that a proposition must be either true or false. This is because in constructive logic, the law of excluded middle must be justified with a proof that is able to demonstrate the truth or falsity of every possible proposition. Against Hilbert, who declared that "each particular mathematical question can be solved in the sense that the question under consideration can either be affirmed, or refuted," Brouwer argued that there were mathematical problems that could not be shown to be true or false. To illustrate this critique, consider Goldbach's conjecture, which states that every even number is the sum of two primes. This is one of the oldest unproved statements in mathematics, and according to intuitionistic logic it must remain so. The reason for this is that while one may demonstrate the truth of Goldbach's conjecture for an arbitrarily large quantity of even numbers ($1 + 3 = 4$, $3 + 3 = 6$, $3 + 5 = 8$, and so on), it would be impossible for any finite entity to demonstrate that this is true for *every* even number, which are infinite. Although no counterexample to Goldbach's conjecture has been demonstrated, neither has a proof been constructed to demonstrate that it is correct. Thus, according to Brouwer, Goldbach's conjecture is neither true nor false (Dalen 2002).

In many ways, the restricted nature of the intuitionist logic pioneered by Brouwer makes it weaker than classical logic, but as it would later be discovered, it has properties that make it ideal for computing. This intersection of intuitionist logic and programming languages is known as the Curry–Howard correspondence. As the mathematicians Haskell Curry and William Howard realized, procedural computer programs and mathematical proofs were the same kind of mathematical object. In other words, since the existence of any object in intuitionistic logic must be proved, this proof can be used as an algorithm to generate that object (Howard 1980).

The importance of Curry and Howard's observations about the isomorphism between these previously separate formal domains—constructive proof systems and systems for computation—can hardly be overstated. The articulation of this correspondence led to a veritable revolution in computing beginning in the 1970s with the development of a new suite of formal systems that comprise at once a proof system and a functional programming language.

Of these new formal systems, one of the best-known and most relevant to interstellar communication was developed by

Thierry Coquand—the calculus of constructions, which has also served as the basis for algorithmic proof checks such as Coq. The calculus of constructions is an extension of the Curry–Howard correspondence insofar as it also incorporates proofs for intuitionistic predicates. In other words, it extends the isomorphism to include quantifiers, which greatly expands the scope of the formal proofs possible within the system.

Another important point worth mentioning is that the Curry–Howard correspondence also sparked a revolution in mathematical linguistics, led primarily by Richard Montague in the 1970s. Whereas Chomsky sought to apply the principles of logic to the syntax of language, Montague sought to do the same with semantics. As Montague wrote in 1970, "there is in my opinion no important theoretical difference between natural languages and the artificial languages of logicians; indeed I consider it possible to comprehend the syntax and semantics of both kinds of languages with a single natural and mathematically precise theory." While teaching introductory logic courses, Montague recognized that the ability to translate a simple sentence in natural language into the language of predicate logic required a bilingual individual—one who understood both the natural language being translated and the logical language. Using the lambda calculus, Montague was the first to develop a mechanical method for translating natural languages into formal languages. In other words, this would obviate the need for the bilingualism—the translation could occur automatically without understanding.

THE ALGORITHMIC TURN

Now that the fundamentals of the lambda calculus and constructive logic have been articulated, let us consider the design of Ollongren's *lingua cosmica*, which will henceforth be referred to as Lincos 2.0. This updated version of Freudenthal's Lincos pulls directly from both formal systems to create a *lingua cosmica* that attempts to satisfy a number of requirements established by Ollongren. The requisite properties of this new system, according to Ollognren, are "linear notation and simple syntax, the self-interpretation of messages, rich message content, redundancy, and the possibility of structuring and sizing the messages" (Ollongren 2013).

As Ollongren rightfully notes, natural languages on Earth are capable of self-interpretation. Young children are able to grasp the fundamentals of their native language without much instruction (exposure and repetition seem to suffice). Only when they are older do they use this basic language to explain the complicated formal nuances (or grammar) of that language. The ability to use a language to explain the same language is what Ollongren means by a language capable of self-interpretation.

In the case of Lincos 2.0, it is obvious that the extraterrestrial will not have recourse to natural human languages sent along with the message as a means of explaining its contents. Thus, the message must be capable of self-interpretation. For this reason, Ollongren eschews a single-level approach like that advocated by Freudenthal, where logic and mathematical signs are all combined in the message to create a rather "unwieldly notation," which was better suited for expressing mathematical relations than conveying information about life on Earth. Instead, Ollongren advocated a multilevel approach for Lincos 2.0. This consists of a basic text that contains the message (comprised of text, images, and other media) and another level of the message, the metalevel, that will provide a means of understanding and interpreting the ground text. This metalevel is essentially an annotation of a text in natural language that is coded using the calculus of constructions with induction.

According to Ollongren, after becoming familiar with the Coq proof-checking system, it became apparent that Coq's underlying formalism could be immensely useful as the basis for a *lingua cosmica*. Coq was developed based on the calculus of constructions, a higher-order typed lambda calculus that functions as both a programming language and as a foundation for mathematics based on constructivism. In fact, Ollongren based Lincos 2.0 on the calculus of constructions with inductions (henceforth CCI), which adds inductive types to the formalism (i.e., functions that are defined in terms of themselves).

CCI allowed Ollongren to meet all the requirements he set for himself in the design of a language for communicating with extraterrestrial intelligences. It is linear with a relatively simple syntax given its small number of primitive elements. It is also capable of describing a wide variety of terrestrial and human experiences by modeling the static logical relations of these experiences using typed lambda expressions. The most important aspect of this language is that it is constructive and deterministic. Since all the objects in the language are constructed, the correctness of the propositions within the system can all be proved within the system. When we consider that an alien recipient of a message written in Lincos 2.0 must begin interpreting the message without context, the fact that the statements are provable and deterministic is of huge importance. The typed lambda calculus, by its design, only allows for the execution of functions of the correct type, specified within the program. This eliminates the possibility of errors or misinterpretation of the message, since the messages would be constructed so that there was only one correct interpretation. Moreover, this construction of the message allows for automatic proof-checking, thus expediting the interpretation of the message and virtually guaranteeing a correct interpretation of its contents.

So what would this look like in practice? We will not dwell too long on concrete examples of Lincos 2.0 as outlined by Ollongren in his monograph, since this text was intended not as a program for transmission but as a compendium of examples for how such a program could be designed. Let us first consider a well-known Aristotelian syllogism that Ollongren provides as a basic example of the notation of his Lincos system:

All humans are mortal, and all Greeks are humans. Thus, all Greeks are mortal.

First, Ollongren defines a universe of discourse D, and introduces humans, mortality, and Greeks as types:

CONSTANT D: Set.CONSTANTS human, mortal, Greek :
D → Prop.

It is worth noting here that Ollongren has slightly adapted the conventional notation of the calculus of constructions for Lincos in order to improve readability, but the concepts conveyed are the same. The declarative items above have defined our environment. Note that "Set" and "Prop" are two distinct predefined types (that is, in an interstellar message, they would have been introduced as types in an earlier example), and D is introduced as an instance of that type. Regarding "human," "mortal," and "Greek," these are of the type D → Prop, which is a mapping function. Thus if some variable x has type D, then the application of the type "human" to x, which is written as (h x) will be of the type "Prop" as a result of the mapping function. The same applies for "mortal" and "Greek" since these are all the same type—namely, mapping functions from type D to type

Prop. So in the Lincos notation, the Aristotelian syllogism would look like this:

(All x : D)((human x) $\rightarrow$ (mortal x)) $\wedge$ (Greek x) $\rightarrow$
　　(human x) $\rightarrow$ (Greek x) $\rightarrow$ (mortal x)

Here, the operators $\rightarrow$ and $\wedge$ retain their denotations from classical logic, namely implication and conjunction. The fact stated above lacks any verification, but verification is a key element of the Lincos system. We will offer a few more examples to demonstrate how facts are verified within this system. The reader should note that a primary feature of the lambda calculus is that it allows for anonymous functions, but this doesn't preclude naming functions in a programming language or Lincos as a matter of convenience. Moreover, when functions in Lincos are defined recursively or inductively (that is, terms defined in terms of themselves), naming conventions must be used. Ollongren demonstrates why this is necessary by offering an inductive definition of factorials for any natural number greater than or equal to zero:

INDUCTIVEfac := $\lambda\ n.(=n\ 0) \rightarrow$ (fac 0) | $\neg(=n\ 0)$
　　$\rightarrow$ (*n (fac($-n$ 1)))

In this inductive definition of a factorial of the natural number n, the vertical stroke separates the two hypotheses from one another. If sent in the message, each of these hypotheses would be evaluated successively depending on the value of n. If the value of n is anything other than 0, the evaluator moves to the right-hand hypothesis and applies the value of n to the function according to the definition. Thus, if we wanted to compute 3! in Lincos notation, we would have to use beta-conversion several times such, which would look like this:

(fac 3) = (*3(fac2) = (*3(*2(fac1))) = (*3(*2(*1(fac0))))
   = (*6(fac0)) = 6

As Ollongren notes, however, despite the prominence of the beta-conversion in typed lambda calculus, its use in Lincos, while not impossible, is limited mostly to expressing processes. Another important point is that the above definition does not use types, which will be necessary for Lincos. To begin, let us assign fac the type nat→nat, indicating that it takes a one natural number as an argument, and returns a natural number (i.e., that fac is of arity 1).

INDUCTIVE fac [λn : nat]: nat := h1 : (= n 0) → (fac 0) |
   h2: ¬(= n 0) → (* n(fac(−n 1)))

Here we see the hypotheses have been named (h1 and h2, respectively), but need types. Thus, we have:

h1: (ALL n : nat)((=n 0) → (fac 0)): nath2: (ALL n : nat)
   (¬(=n 0) → (* n(fac(−n 1)))): nat

Also note that here we have introduced the universal quantifier, ALL, which ranges over variables in the lambda expression in the same way it would in classical predicate logic. A final aspect of Lincos that Ollongren introduces is "Set," which is not to be taken for its strict set theoretical meaning with its accompanying operators such as intersection or union, but rather as a "basic collection of entities" grouped together. Ollongren introduces this notion as a way of explaining type checking, which is essentially the formal method of substituting entities of the same type. Consider his definitions:

DEFINE I := [λ x: Set].x DEFINE K := [λ c, x : Set].c

As Ollongren notes, this results in

I: Set → Set K: Set → Set → Set

I is an identity function because for any x : Set, I x : Set and I $x=x$. Likewise, any c, d: Set application of the K function results in K c, d : Set, and K c d = c. In other words, both the I and the K functions are identity functions—this means they type check and are functionally equivalent. Therefore Ollongren refers to this operation as the "chameleon effect," since it is like the chameleon's ability to change its color while retaining its identity.

Now, let us consider a relatively simple example of using the Lincos system to annotate a sentence describing a static situation, such as a book lying upon a shelf. The example Ollongren provides is *"Alice in Wonderland* is a book. It lies on some shelf number 1."* First, the environment must be defined:

CONSTANTS book, shelf : Set

CONSTANT *Alice in Wonderland* : book

CONSTANT #1 : shelfCONSTANT is-book : book → Prop

CONSTANT is-shelf : shelf → Prop.

Next we will define a function, namely the function of a book lying on a shelf (BS). The lambda abstraction demonstrates that the BS function requires two arguments:

DEFINE BS : book → shelf → Prop := [λ x :book; y : shelf]
(is-book x) ∧ (is-shelf y)

The reason for the BS function requiring two arguments is because:

(x : book) (is-book x) : Prop(x : shelf) (is-shelf x) :
Prop[λ x:book; y:shelf] (is-book x) : book → shelf → Prop

Notice that the variables x and y are not explicitly introduced with a declarator but introduced implicitly in the context of the lambda function. In this sense they are locally bound to this lambda abstraction, meaning that these variables need not be of the same type if they appear elsewhere in the message. Now, we have *Alice in Wonderland* as a book, and no. 1 as a shelf. Thus:

(BS *Alice in Wonderland* #1) : Prop

which implies

(BS *Alice in Wonderland* #1) → (is-book *Alice in Wonderland*)
  ∧ (is-shelf #1)

This allows us to construct an entity (which Ollongren has named f.2.1) that has this type:

FACT f.2.1 : (BS *Alice in Wonderland* #1) → (is-book *Alice in Wonderland*) ∧ (is-shelf #1)

Note that this object in this form cannot be a declarator, since facts in the constructive calculus (and by extension, Lincos) also require verification. The advantage of Lincos 2.0 is that, assuming the environment to be sound, all the facts communicated therein are *guaranteed to be correct*. Moreover, the proof of their correctness can be provided within Lincos itself. To understand this crucial interplay between the environment and facts, Ollongren turns to the philosopher Ludwig Wittgenstein, who was also concerned with precisely this relationship—namely, the conditions on which one fact could be said to represent another. Or put differently, the relationship between statements in language and states of affairs in the world. As Wittgenstein argues in the *Tractatus* (1922), under the heading of his second thesis— "What is the case, the fact, is the existence of atomic facts"—is that the arrangement of the atomic facts (what he terms objects) "produces states and affairs." These states of affairs stand in a definite relation to one another and the "totality of existing states of affairs is the world." In the case of Lincos, Ollongren conceives of the state of affairs created by objects as the Lincos environment (that is, the defined terms) and the relationship between these objects (facts) is articulated through the conventions of the calculus of constructions. Thus, a fact is in a sense structured, and this structure is determined by the proof that is used to create the fact (object).

Now let us consider the receipt of the above syllogism by an alien recipient. In Ollongren's formulation, this syllogism would be included as a basic text of an interstellar message written in a natural language—in this case, English. At the metalevel, the logical form of basic text would be shown and accompanied by its proof. But how is the alien recipient to understand any of this? According to Ollongren (2013), it is "not unreasonable to assume that an intelligent species receiving an interstellar message unmistakably bearing a linguistic signature will put automatic information processing machines at work to do the decoding and some of the interpretation." After all, we would likely do the same thing if we were to receive an extraterrestrial message on Earth. This automatic processing machine would likely recognize that the meta-message made use of constructive logic. Assuming there is a lot of redundancy in the message, this would allow the extraterrestrial recipient to attribute the sign "ALL" to its concept of universal quantification, the sign → to implication, and ∧ to conjunction. Moreover, since the terms "human," "mortal," and "Greek" appear in both the metalevel text and the basic natural language text, this will provide the stepping stone for moving from logical language to natural language interpretations. As Ollongren notes, however, the assumption that extraterrestrials would understand any part of the message conveyed in Lincos 2.0 means that there would have to be a lot of redundancy to the message—several variations of this

Aristotelian syllogism ought to be included in the message. As for distinguishing the metalevel from the basic level, Ollongren proposes "using a kind of characteristic delimiters" to separate the levels during the actual transmission.

As Ollongren points out, it is conceivable to develop some type of functional programming language as a computer implementation of the system he described in *Astrolinguistics*. This would be a Coq-like proof-checker that could be deployed not only in the design of the message (indeed, many of the examples in the text were confirmed with a proof-checker), but this verification machinery could in fact be provided to the extraterrestrials at some level beyond the metalevel annotation. Ollongren has helpfully left this for others to work out the details.

At its core, Lincos 2.0 is an effort to provide an annotation to a natural language (or visual message) used in interstellar communication. The hope is that by roughly articulating the logical syntax of natural language, this will aid in the interpretation of the natural language message. Of course, the semantic content of this message may be impossible for the extraterrestrial to reach. Consider the basic example of *"Alice in Wonderland* is a book. It lies on some shelf no. 1." Lincos is able to articulate the relationship between two objects—*Alice in Wonderland* and a shelf no. 1—but an extraterrestrial recipient is unlikely to have read Lewis Carroll's masterpiece, so such an assertion would not mean much. Of course, the examples provided in Ollongren's monograph are not intended to be the actual contents of an interstellar message—there'd be little point in discussing the spatial arrangements of children's books with aliens. To establish a connection between logical syntax of a text and its semantic content, it will likely be necessary to begin with mathematical or scientific notions that may be presumed to be universal (e.g., the relationship between a circle's circumference and diameter, the properties of hydrogen, or aspects of certain celestial objects). The exchange of meaningful information may itself be abetted with the use of extralinguistic means, whether this is still photos, video, or even music. These "texts" may also be annotated in Lincos to explain how they work—that is, that a video signal is an array of so many pixels, which are organized into a still frame, the relationship between one frame to the next, the speed of transition between the frames, and so forth. It is not difficult to imagine explicitly defining the rules for reconstructing an image using Lincos. Ollongren even argues that extralinguistic cues could be used as a second metalevel helping to elucidate the purpose of the first metalevel, which is coded in Lincos.

In the absence of some multilevel system, however, the question remains whether Ollongren's two-level system would be able to facilitate the understanding of some basic level text written in natural language. Until we have established contact with extraterrestrials, such questions remain speculative. Nevertheless, it is possible to see whether Lincos could be used to effectively communicate on Earth in the absence of knowledge about some basic test. In 2000, Ollongren carried out such a test with his graduate student Johanna Novozamsky. Novozamsky, fluent in Czech, created the formal annotations for seventy-four sentences written in Czech; the logical relations between the elements of the sentence were then coded using the French Coq system and given to Ollongren, who knew neither the content of the sentences nor the lexicon/grammar of the Czech language.

Coq is not a perfect analogue to Lincos, but both are based on the calculus of constructions with induction and it thus offers a nice comparison. The experiment was also burdened by the fact that both Ollongren and Novozamsky were well acquainted with constructive logic. In the first place, Ollongren noted that the annotated version of the Czech text was far larger than the text itself, since the annotations were also accompanied by their proofs. This will also be an aspect of any Lincos message. Moreover, after comparing his interpretation with the actual meaning of the text, Ollongren found that his "interpretation was far, but indeed not too far off the mark!" This suggests that "any informative text meant for transmission to ETI must be composed with extreme care so as to ensure as little freedom of interpretation as possible." A way to limit misunderstandings is to limit the number of actors and actions in a given transmission. The experiment was not entirely successful, but did suggest that "much understanding of textual contents could be achieved—but the text was far from completely understood" (Ollongren 2013).

REFERENCES

Almár, I., and H. Shuch. 2007. The San Marino scale: A new analytical tool for assessing transmission risk. *Acta Astronautica* 60:57–59.

Aristotle. 1912. *Metaphysics*. Trans. W. D. Ross. Oxford: Clarendon Press. Original work 350 BCE.

Arbib, M. 1979. Minds and millenia: The psychology of interstellar communication. *Cosmic Search* 1:21.

Atwell, E., and J. Elliott. 2001. A corpus for interstellar communication. In *Proceedings of CL2001: International Conference on Corpus Linguistics*, ed. P. Rayson, A. Wilson, T. McEnery, A. Hardie and S. Khoja, 31–39. Lancaster: University Centre for Computer Corpus Research on Language Technical Papers.

Atwell, E., and J. Elliott. 2002. Corpus linguistics and the design of a response message. Paper presented at the 53rd International Astronautical Congress, Houston, Texas.

Backus, J., E. L. Bauer, J. Green, C. Katz, J. McCarthy, A. J. Perlis, et al. 1963. Revised report on the algorithm language ALGOL 60. *Communications of the ACM* 6 (1): 1–17.

Bainbridge, W. 2013. Direct contact with extraterrestrials via computer emulation. In *Civilizations Beyond Earth: Extraterrestrial Life and Society*, ed. D. Vakoch. New York: Berghahn.

Balaguer, M. 1998. *Platonism and Anti-Platonism in Mathematics*. Oxford: Oxford University Press.

Ball, R. 1901. Signaling to Mars. *Scientific American Supplement*, June 8, 212.

Bania, T., and R. Rood. 1993. Search for interstellar beacons at the $^3He^+$ HY-perfine transition frequency. In *Third Decennial US–USSR Conference on SETI*, ed. S. Shostak. San Francisco: Astronomical Society of the Pacific.

Bassi, B. 1992. Were it perfect, would it work better? Survey of a language for cosmic intercourse. *Versus* 61/63.

Benford, G., J. Benford, and D. Benford. 2010a. Messaging with cost-optimized interstellar beacons. *Astrobiology* 10 (5): 475–490.

Benford, G., J. Benford, and D. Benford. 2010b. Searching for cost-optimized interstellar beacons. *Astrobiology* 10 (5): 491–498.

Berwick, R., and N. Chomsky. 2016. *Why Only Us? Language and Evolution*. Cambridge, MA: MIT Press.

Billingham, J., and J. Benford. 2011. Costs and difficulties of large-scale "messaging," and the need for international debate on potential risks. *arXiv*:1102.1938.

Billings, L. 2016. $100-million plan to send probes to the nearest star. *Scientific American*, April 12.

Blair, D., and M. Zadnik. 1993. A list of possible interstellar communication channel frequencies for SETI. *Astronomy and Astrophysics* 278: 669–672.

Blum, H. 1990. SETI, phone home. *New York Times Magazine*, October 21.

Boole, G. 2003. *An Investigation of the Laws of Thought on Which Are Founded the Mathematical Theories of Logic and Probabilities*. Amherst, NY: Prometheus Books. Originally published 1854.

Bornkessel-Schlesewsky, I., M. Schlesewsky, S. Small, and J. Rauschecker. 2015. Neurobiological roots of language in primate audition: Common computational properties. *Trends in Cognitive Sciences* 19: 142–150.

Boroson, D., J. Scozzafava, D. Murphy, B. Robinson, B., and H. Shaw. 2009. The lunar laser communications demonstration (LLCD). Paper presented at the 3rd IEE International Conference on Space Mission Challenges for Information Technology, Washington, DC, July 2009.

Braastad, R., and A. Zaitsev. 2003. Synthesis and transmission of Cosmic Call 2003 interstellar radio message. The Institute of Radio Engineering and Electronics of RAS. http://www.cplire.ru/html/ra&sr/irm/CosmicCall-2003/index.html.

Brin, D. 2014. The search for extraterrestrial intelligence (SETI) and whether to send "messages" (METI): A case for conversation, patience, and due diligence. *Journal of the British Interplanetary Society* 67:8–16.

Brinkmann, H., L. Commare, H. Leder, and R. Rosenberg. 2014. Abstract art as a universal language? *Leonardo* 47:256–257.

Brouwer, L. E. J. 1981. *Brouwer's Cambridge Lecture on Intuitionism*. Cambridge: Cambridge University Press. Originally delivered at Cambridge University 1951.

Brouwer, L. E. J. 1996. Mathematics, science and language. In *From Kant to Hilbert: A Source Book in the Foundations of Mathematics*, vol. 2, ed. W. Ewald, 1170–1185. Oxford: Oxford University Press. Originally published 1928.

Brown, M. 2005. Radio Mars: The transformation of Marconi's popular image, 1919–1922. In *Transmitting the Past: Historical Perspectives on Broadcasting*, ed. J. Winn and S. Brinson, 16–32. Tuscaloosa, AL: University of Alabama Press.

Busch, M., and R. Reddick. 2011. Testing SETI message designs. In *Communication with Extraterrestrial Intelligence (CETI)*, ed. D. Vakoch. Albany, NY: SUNY Press.

Bush, V. 1945. *Science, the Endless Frontier: A Report to the President*. Washington, DC: US Office of Scientific Research and Development.

Cameron, A. 1963. *Interstellar Communication*. Minneapolis: University of Minnesota Press.

Camille Flammarion's latest views on Martian signaling. 1907. *Scientific American Supplement* 1652, August 31, 137.

Cerceau, F. 2015. Fraction of civilizations that develop a technology that releases detectable signs of their existence into space, fc, pre-1961. In *The Drake Equation: Estimating the Prevalence of Extraterrestrial Life through the Ages*, ed. D. Vakoch and M. Dowd, 205–225. Oxford: Oxford University Press.

Chadwick, J. 1990. *The Decipherment of Linear B*. Cambridge: Cambridge University Press. Originally published 1967.

Chomsky, N. 1957. *Syntactic Structures*. The Hague: Mouton.

Chomsky, N. 2000. Minimalist inquiries: The framework. In *Step by Step: Essays on Minimalist Syntax in Honor of Howard Lasnik*, ed. R. Martin, D. Michaels, and J. Uriagereka, 89–155. Cambridge, MA: MIT Press.

Chomsky, N. 2002. *On Nature and Language*. Cambridge: Cambridge University Press.

Chomsky, N., and J. Gliedman. 1983. Things no amount of learning can teach. *Omni* 6 (11). https://chomsky.info/198311__/.

Church, A. 1932. A set of postulates for the foundation of logic. *Annals of Mathematics* 33:346–366.

Church, A. 1936. An unsolvable problem of elementary number theory. *American Journal of Mathematics* 58 (2): 345–363.

Cocconi, G., and P. Morrison. 1959. Searching for interstellar communications. *Nature* 184:844–846.

Cockell, C. 2018. *The Equations of Life: How Physics Shapes Evolution*. London: Atlantic Books.

Coe, K., C. Palmer, and C. Pomianek. 2013. ET phone Darwin: What can an evolutionary understanding of animal communication and art contribute to our understanding of methods for interstellar communication?

In *Civilizations Beyond Earth: Extraterrestrial Life and Society*, ed. D. Vakoch and A. Harrison, 214–225. New York: Berghahn.

Copple, K. 2008. Bringing AI to life. In *Parsing the Turing Test*, ed. R. Epstein, G. Roberts, and G. Beber. New York: Springer.

Cordes, J., and T. Lazio. 1991. Interstellar scattering effects on the detection of narrow-band signals. *Astrophysical Journal* 376:123–133.

Corum, K., and J. Corum. 2003. *Nikola Tesla and the Planetary Radio Signals*. Adapted from a paper presented at the 5th International Tesla Conference: Tesla III Millennium, 1996, Belgrade, Serbia.

Crowe, M. 1999. *The Extraterrestrial Life Debate: 1750–1900*. Mineola, NY: Dover.

Dalen, D. 2002. Intuitionistic logic. In *Handbook of Philosophical Logic*, vol. 5, ed. D. Gabbay and F. Guenthner, 1–114. New York: Springer.

Darwin, C. 2003. *On the Origin of the Species*. New York: Signet. Originally published 1859.

Davies, N. 1967. Bishop Godwin's "lunatique language." *Journal of the Warburg and Courtauld Institutes* 30:296–316.

Dehaene, S. 2011. *The Number Sense: How the Mind Creates Mathematics*. Rev. and exp. ed. Oxford: Oxford University Press.

DeVito, C. 2010. Alien mathematics. *Journal of the British Interplanetary Society* 63 (8): 306–309.

DeVito, C. 2011a. Cultural aspects of interstellar communication. In *Civilizations Beyond Earth*, ed. D. Vakoch and A. Harrison, 159–169. New York: Berghahn.

DeVito, C. 2011b. On the universality of human mathematics. In *Communications with Extraterrestrial Intelligence*, ed. D. Vakoch, 439–488. Albany, NY: SUNY Press.

DeVito, C., and R. Oehrle. 1990. A language based on the fundamental facts of science. *Journal of the British Interplanetary Society* 43 (12): 561–568.

Dick, S. 2006. The postbiological universe. Paper presented at the 57th International Astronautical Congress, Valencia, Spain, October 2006.

Doyle, L., B. McCowan, S. Johnston, and S. Hanser. 2011. Information theory, animal communication, and the search for extraterrestrial intelligence. *Acta Astronautica* 68 (3–4): 406–417.

Drake, F., and G. Helou. 1977. The optimum frequencies for interstellar communications as influenced by minimum bandwidth. Unpublished NAIC report.

Drake, F., and D. Sobel. 1992. *Is Anyone Out There?* New York: Delacorte Press.

Dumas, S. N.d. The 1999 and 2003 messages explained. https://www .plover.com/misc/Dumas-Dutil/messages.pdf.

Dumusque, X., F. Pepe, C. Lovis, D. Ségransan, J. Sahlmann, W. Benz, et al. 2012. An Earth-mass planet orbiting α Centauri B. *Nature* 491:207–211.

Dutil, Y., and S. Dumas. 2016. Annotated Cosmic Call primer. *Smithsonian*, September. https://www.smithsonianmag.com/science-nature/ annotated-cosmic-call-primer-180960566/.

Dutil, Y., and S. Dumas. 2001. Error correction scheme in active SETI. Paper presented at 52nd International Astronautical Congress, Toulouse, France, October 2001.

Džamonja, M. 2017. Set theory and its place in the foundations of mathematics: A new look at an old question. *Journal of Indian Council of Philosophical Research* 34:415–424.

Eco, U. 1995. *The Search for the Perfect Language.* Hoboken, NJ: Wiley.

Einstein, A. 1922. *Sidelights on Relativity.* London: Methuen & Co.

Ekers, R., D. Cullers, J. Billingham, and L. Scheffer, eds. 2002. *SETI 2020: A Roadmap for the Search for Extraterrestrial Intelligence.* Mountain View, CA: SETI Press.

Elliott, J. 2011a. A semantic "engine" for universal translation. *Acta Astronautica* 68 (3–4): 435–440.

Elliott, J. 2011b. A human language corpus for interstellar message construction. *Acta Astronautica* 68 (3-4): 418-424.

Elliott, J., E. Atwell, and B. Whyte. 2000a. Increasing our ignorance of language: Identifying language structure in an unknown "signal." In *Proceedings of CoNLL-2000 and LLL-2000*, ed. C. Cardie, W. Daelemans, C. Nédellec and E. Sang, 25–30. Lisbon: Association for Computational Linguistics.

Elliott, J., E. Atwell, and B. Whyte. 2000b. Language identification in unknown signals. In *COLING '00 Proceedings of the 18th Conference on Computational Linguistics*, 1021–1015. San Francisco: Morgan Kaufmann.

Embick, D., A. Marantz, Y. Miyashita, W. O'Neil, and K. Sakai. 2000. A syntactic specialization for Broca's area. *Proceedings of the National Academy of Sciences* 97:6150–6154.

Ethergrams to Mars. 1906. *Electrical Record*, June 13.

Everett, D. 2005. Cultural constraints on grammar and cognition in Pirahã. *Current Anthropology* 46 (4): 621–646.

Fitzpatrick, P. 2003. Outline of Cosmic OS. https://people.csail.mit.edu/paulfitz/cosmicos.shtml.

Fitzpatrick, P. 2014. Cosmic OS: A coder's contact message. https://cosmicos.github.io/.

Fleming, G. 1909. Signalling to Mars with mirrors. *Scientific American* 100 (22): 407.

Fleury, B. 1980. The aliens in our oceans: Dolphins as analogs. *Cosmic Search* 6:2–5.

Freudenthal, H. 1960. *Lincos: Design of a Language for Cosmic Intercourse, Part 1*. Amsterdam: North-Holland.

Galileo, G. 1957. *The Assayer*. In *Discoveries and Opinions of Galileo*, trans. D. Stillman. New York: Doubleday. Originally published 1623.

Galileo, G. 2001. *Dialogue Concerning the Two Chief World Systems: Ptolemaic and Copernican*. Trans. S. Drake and S. Gould. New York: Modern Library. Originally published 1632.

Galton, F. 1896. Intelligible signals between neighboring stars. *Fortnightly Review* 60:657–664.

Garber, S. 1999. Searching for good science: The cancelation of NASA's SETI program. *Journal of the British Interplanetary Society* 52:3–12.

Garrett, M., A. Siemion, and W. van Cappellen. 2016. *All-sky radio SETI*. Paper presented at the MeerKAT Science: On the Pathway to SKA workshop, Stellenbosch, South Africa, May.

Gertz, J. 2016. Reviewing METI: A critical analysis of the arguments. *Journal of the British Interplanetary Society* 69:31–36.

Gödel, K. 1992. *On Formally Undecidable Propositions of Principia Mathematica and Related Systems*. Trans. B. Meltzer. New York: Dover. Originally published in German 1931.

Gold, T. and J. Pumphrey. 1948. Hearing. I. The cochlea as a frequency analyzer. *Proceedings of the Royal Society B* 135 (881): 462–491.

Gombrich, E. 1972. The visual image. *Scientific American* 227:82–97.

Gravemeijer, K., and J. Terwel. 2000. Hans Freudenthal: A mathematician on didactics and curriculum theory. *Journal of Curriculum Studies* 32 (6): 777–796.

Guillochon, J., and A. Loeb. 2015. SETI via leakage from light sails in exoplanetary systems. *Astrophysical Journal Letters* 811 (2): 1–6.

Haqq-Misra, J., M. Busch, S. Som, and S. Baum. 2013. The benefits and harms of transmitting into space. *Space Policy* 29: 40–48.

Hauser, M., N. Chomsky, and T. Fitch. 2002. The faculty of language: What is it, who has it, and how did it evolve? *Science* 298:1569–1579.

Hello, Mars—this is the Earth! 1919. *Popular Science Monthly*, September, 74–75.

Herman, L. 2010. What laboratory research has told us about dolphin cognition. *International Journal of Comparative Psychology* 23: 310–330.

Herman, L., and P. Forestell. 1985. Reporting presence or absence of named objects by a language-trained dolphin. *Neuroscience and Biobehavioral Reviews* 9:667–681.

Herman, L., D. Richards, and J. Wolz. 1984. Comprehension of sentences by bottlenosed dolphins. *Cognition* 16: 129-219.

Herzing, D. 2010. SETI meets a social intelligence: Dolphins as a model for real-time interaction and communication with a sentient species. *Acta Astronautica* 67 (11–12): 1451–1454.

Herzing, D. 2014. CHAT: Is it a dolphin translator or an interface? http://www.wilddolphinproject.org/chat-is-it-a-dolphin-translator-or-an-interface/.

Highwater, J. 1983. *The Primal Mind: Vision and Reality in Indian America.* New York: Book Sales.

Hilbert, D. 1928. The foundations of mathematics. In *The Emergence of Logical Empiricism: From 1900 to the Vienna Circle*, ed. S. Sarkar. New York: Garland, 1996.

Hippke, M., and J. Learned. 2018. Interstellar communication IX. Message decontamination is possible. *axXiv*:1802.02180.

Hockett, C. 1966. The problem of universals in language. In *Universals in Language*, ed. J. Greenberg. Cambridge, MA: MIT Press.

Hogben, L. 1952. Astraglossa, or First steps in celestial syntax. *Journal of the British Interplanetary Society* 9:258–274.

Hogben, L. 1961. Cosmical language. *Nature* 192 (4805): 826–827.

Horner, F. 1957. Radio noise from planets. *Nature* 180:1253.

Horowitz, P. and C. Sagan. 1993. Five years of Project META: An all-sky narrow-band radio search for extraterrestrial signals. *Astrophysical Journal* 415:218–235.

Howard, W. 1980. The formulae-as-types notion of construction. In J. *To H.B. Curry: Essays on Combinatory Logic, Lambda Calculus and Formalism*, ed. Seldin and J. Hindley, 479–490. Cambridge, MA: Academic Press.

International Academy of Astronautics. 2007. *Sending Communications to Extraterrestrial Civilizations.* https://iaaseti.org/en/protocols/.

Inter-stellar wireless a possibility to Marconi. 1919. *The Wireless Age*, March, 10–11.

Janik, V. 2000. Whistle matching in wild bottlenose dolphins. *Science* 289:1355–1357.

Jensen, H. 1998. *Self-Organized Criticality.* Cambridge: Cambridge University Press.

Jerison, H. 1985. Animal intelligence as encephalization. *Philosophical Transactions of the Royal Society B* 308:21–35.

Johnson, P. 1972. The genesis and development of set theory. *Two-Year College Mathematics Journal* 3:55–62.

Johnson, S., W. Straten, M. Kramer, and M. Bailes. 2001. High time resolution observations of the Vela pulsa. *Astrophysical Journal* 549: L101–104.

Johnson, M. 2009. How the statistical revolution changes (computational) linguistics. In *Proceedings of the EACL 2009 Workshop on the Interaction between Linguistics and Computational Linguistics*, 3–11. Athens: Tehnografia Digital Press.

Kaiho, K., and N. Oshima. 2017. Site of asteroid impact changed the history of life on Earth: The low probability of mass extinction. *Scientific Reports* 7:1–12.

Kaiser, A. 2004. Sound as intercultural communication: A meta-analysis of music with implications for SETI. *Leonardo* 37:36–37.

Kardashev, N. 1979. Optimal wavelength region for communication with extraterrestrial intelligence: $\lambda = 1.5$mm. *Nature* 278:28–30.

Kleene, S., and J. Rosser. 1935. The inconsistency of certain formal logics. *Annals of Mathematics* 36 (3): 630–636.

Lakoff, G., and N. Nuñez. 2000. *Where Mathematics Comes From: How the Embodied Mind Brings Mathematics into Being*. New York: Basic Books.

Lee, R., P. Jonathan and P. Ziman. 2010. Pictish symbols revealed as a written language through the application of Shannon entropy. *Proceedings of the Royal Society A* 466:2545–2560.

Lemarchand, G., and J. Lomberg. 2009. Universal cognitive maps and the search for intelligent life in the universe. *Leonardo* 42:396–402.

Letaw, H. 2005, October. A study on interstellar storytelling. Paper presented at the 56th International Astronautical Congress, Fukuoka, Japan.

Letaw, H. 2013. Cosmic storytelling: Primitive observables as Rosetta analogies. In *Civilizations Beyond Earth: Extraterrestrial Life and Society*, ed. D. Vakoch and A. Harrison, 170–190. New York: Berghahn Books.

Let the stars alone. 1919. *New York Times*, January 21.

Lilly, J. 1962. *Man and Dolphin*. Worthing: Littlehampton Book Services.

Lilly, J. 1969. *The Mind of the Dolphin: A Nonhuman Intelligence*. New York: Avon.

Liu, Z., Y. Cai, Y. Wang, Y. Nie, C. Zhang, Y. Xu, et al. 2018. Cloning of macaque monkeys by somatic cell nuclear transfer. *Cell* 172:881–887.

Lyn, H., J. Russell, D. Leavens, K. Bard, S. Boysen, J. Schaeffer, and D. Hopkins. 2014. Apes communicate about absent and displaced objects: Methodology matters. *Animal Cognition* 17:85–94.

Malyshev, D., K. Dhami, T. Lavergne, T. Chen, N. Dai, J. Foster, I. Correa, and F. Romesberg. 2014. A semi-synthetic organism with an expanded genetic alphabet. *Nature* 509:385–388.

Manger, P. 2013. Questioning the interpretations of behavioral observations of cetaceans: Is there really support for a special intellectual status for this mammalian order? *Neuroscience* 250:664–696.

Marconi sure Mars flashes messages. 1921. *New York Times*, September 2.

Marino, L., R. Connor, E. Fordyce, L. Herman, P. Hof, L. Lefebvre, et al. 2007. Cetaceans have complex brains for complex cognition. *PLoS Biology* 5.

Martin-Löf, P. 1996. On the meanings of the logical constants and the justifications of the logical laws. *Nordic Journal of Philosophical Logic* 1 (1): 11–60.

Mayr, E. 1985. The probability of extraterrestrial life. In *Extraterrestrials: Science and Alien Intelligence*, ed. E. Regis. Cambridge: Cambridge University Press.

McCarthy, J. 1963. A basis for a mathematical theory of computation. *Studies in Logic and the Foundations of Mathematics* 35:33–70.

McCarthy, J. 1974. Possible forms of intelligence: natural and artificial. In *Interstellar Communication: Scientific Perspectives*, ed. C. Ponnamperuma and A. G. W. Cameron, 79–87. Boston: Houghton Mifflin.

McConnell, B. 2001. *Beyond Contact: A Guide to SETI and Communicating with Alien Civilizations*. Sebastopol, CA: O'Reilly.

McCowan, B., L. Doyle, and S. Hanser. 2002. Using information theory to assess the diversity, complexity, and development of communicative repertoires. *Journal of Comparative Psychology* 116 (2): 166–172.

McCowan, B., S. Hanser, and Doyle, L. 1999. Quantitative tools for comparing animal communication systems: Information theory applied to bottlenose dolphin whistle repertoires. *Animal Behaviour* 57:409–419.

Mercier, A. 1899. *Communications Avec Mars*. Paris: Astronomical Society of France.

Michaud, M. 2005. Active SETI is not scientific research. SETILeague.org, February. http://www.setileague.org/editor/actvseti.htm.

Minsky, M. 1985. Communication with alien intelligence. *Byte Magazine* 10 (4): 126–142.

Mitchell, J. 2002. *Concepts in Programming Languages*. Cambridge: Cambridge University Press.

Mme. Guzman's curious will. 1891. *Chicago Tribune*, September 14.

Montague, R. 1970. Universal grammar. *Theoria* 36:373–398.

Montemurro, M., and D. Zanette. 2011. Universal entropy of word ordering across linguistic families. *PLoS One* 6.

Monti, M., L. Parsons, and D. Osherson. 2009. The boundaries of language and thought in deductive inference. *Proceedings of the National Academy of Sciences* 106:12554–12559.

Moore, P. 2000. *The Data Book of Astronomy*. Boca Raton, FL: CRC Press.

Moro, A., M. Tettamanti, D. Perani, C. Donati, S.F. Cappa, and F. Fazio. 2001. Syntax and the brain: Disentangling grammar by selective anomalies. *NeuroImage* 13:110–118.

Moro, A. 2013. *The Equilibrium of Human Syntax: Symmetries in the Brain*. New York: Routledge.

Moro, A. 2016. *Impossible Languages*. Cambridge, MA: MIT Press.

Morris, S. 2003. *Life's Solution: Inevitable Humans in a Lonely Universe*. Cambridge: Cambridge University Press.

Musso, P. 2012. The problem of active SETI: An overview. *Acta Astronautica* 78:43–54.

Nieman, H., and C. Nieman. 1920. What shall we say to Mars? *Scientific American* 112 (12): 298.

No Mars message yet, Marconi radios; ends yacht trip "listening in" on planet today. 1922. *New York Times*, June 16.

Ogden, C., and I. Richards. 1989. *The Meaning of Meaning: A Study of the Influence of Language upon Thought and of the Science of Symbolism*. San Diego, CA: Harcourt Brace Jovanovich. Originally published 1923.

Oliver, B., and J. Billingham. 1972. *Project Cyclops: A Design Study of a System for Detecting Extraterrestrial Intelligent Life* (NASA-CR-114445). Washington, DC: NASA.

Ollongren, A. 2004. Large-size message construction for ETI. *Leonardo* 37 (1): 38–39.

Ollengren, A. 2011a. Aristotelian syllogisms. *Acta Astronautica* 68 (3–4): 549–553.

Ollongren, A. 2011b. Large-size message construction for ETI: Inductive self-interpretation in LINCOS. *Acta Astronautica* 68 (3–4): 539–543.

Ollongren, A. 2011c. Recursivity in Lingua Cosmica. *Acta Astronautica* 68 (3–4): 544–548.

Ollongren, A. 2013. *Astrolinguistics: Design of a Linguistic System for Interstellar Communication Based on Logic*. New York: Springer.

Ollongren, A., and D. Vakoch. 2011. Typing logic contents using Linguia Cosmica. *Acta Astronautica* 68 (3–4): 535–538.

Patel, A. 2003. Language, music, syntax and the brain. *Nature* 6 (7): 674–681.

Patterson, F., and R. Cohn. 1990. Language acquisition in lowland gorillas: Koko's first ten years of vocabulary development. *Word* 41:97–143.

Pešek, R. 1979. Activities of the IAA CETI committee from 1965–1976 and CETI outlook. *Acta Astronautica* 6 (1–2): 3–9.

Pickering, W. 1909. Signaling to Mars. *Scientific American* 101 (3): 43.

Popper, K. 1959. *The Logic of Scientific Discovery*. New York: Basic Books.

Quine, W. V. 1960. *Word and Object*. Cambridge, MA: MIT Press.

Radio to the stars, Marconi's hope. 1919. *New York Times*, January 20.

Rajpaul, V., S. Aigrain, and S. Roberts. 2015. Ghost in the time series: No planet for Alpha Cen B. *Monthly Notices of the Royal Astronomical Society* 456:L6–L10.

Raulin-Cerceau, F. 2010. The pioneers of interplanetary communication: From Gauss to Tesla. *Acta Astronautica* 67 (11–12): 1391–1398.

Reddy, F. 2011. *Celestial Delights*. New York: Springer.

Richards, D., J. Wolz, and L. Herman. 1984. Vocal mimicry of computer-generated sounds and vocal labeling of objects by a bottlenosed dolphin, *Tursiops truncatus. Journal of Comparative Psychology* 98 (1): 10–28.

Ridpath, I. 1978. A signaling strategy for interstellar communication. *Journal of the British Interplanetary Society* 31 (3): 108–109.

Rigge, W. 1920. Wireless signals to Mars. *Popular Astronomy* 28:306.

Rose, C., and G. Wright. 2004. Inscribed matter as an energy-efficient means of communication with an extraterrestrial civilization. *Nature* 431:47–49.

Ross, M., and T. Curran. 1965. Optimum detection thresholds in optical communications. *Proceedings of the IEEE* 53:1770–1771.

Roth, G. 2015. Convergent evolution of complex brains and high intelligence. *Philosophical Transactions of the Royal Society B* 370 (1684): 1–9.

Russell, B., and A. Whitehead. 1925. *Principia Mathematica*, vol. 1. Cambridge: Cambridge University Press.

Sagan, C. 1973. *Communication with Extraterrestrial Intelligence*. Cambridge, MA: MIT Press.

Sagan, C. 1985. *Contact*. New York: Simon and Schuster.

Sagan, C., F. Drake, A. Druyan, T. Ferris, J. Lomberg, and L. Sagan. 1978. *Murmurs of Earth: The Voyager Interstellar Record*. New York: Ballantine Books.

Sagan, C., L. Sagan, and F. Drake. 1972. A message from Earth. *Science* 175 (4024): 881–884.

Sandberg, A., S. Armstrong, and M. Cirkovic. 2017. That is not dead which can eternal lie: The aestivation hypothesis for resolving Fermi's paradox. *arXiv*:1705.03394.

Saussure, F. 2011. *Course in General Linguistics*. Trans. Wade Baskin. New York: Columbia University Press. Originally published 1916.

Schwartz, R. N., and C. H. Townes. 1961. Interstellar and interplanetary communication by optical masers. *Nature* 190:205–208.

Secor, H. 1920. Hello Mars! *Electrical Experimenter* 7:1248–1250, 1302, 1304, 1306–1309.

Shannon, C. 1948. A mathematical theory of communication. *Bell System Technical Journal* 27: 379–423.

Shannon, C., and W. Weaver. 1949. *The Mathematical Theory of Communication.* Urbana: University of Illinois Press.

Shklovskii, I., and C. Sagan. 1966. *Intelligent Life in the Universe.* New York: Dell.

Shostak, S. 1995. SETI at wider bandwidths? In *Progress in the Search for Extraterrestrial Life: 1993 Bioastronomy Symposium*, 447–454, ed. S. Shostak. San Francisco: Royal Astronomical Society of the Pacific.

Shostak, S. 2009. When will we find the extraterrestrials? *Engineering and Science* (Spring): 12–21.

Shostak, S. 2010. What ET will look like and why should we care. *Acta Astronautica* 67 (9–10): 1025–1029.

Shostak, S. 2011. Limits on interstellar messages. *Acta Astronautica* 68 (3): 366–371.

Shostak, S. 2013. Are transmissions to space dangerous? *International Journal of Astrobiology* 12:17–20.

Shuch, H. P., and I. Almár. 2007. Shouting in the jungle: The SETI transmission debate. *Journal of the British Interplanetary Society* 60:142–146.

Silver, D., J. Schrittwieser, K. Simonyan, I. Antonoglou, A. Huang, A. Guez, et al. 2017. Mastering the game of Go without human knowledge. *Nature* 550:354–359.

Silverberg, R. 1974. Schwartz between galaxies. In *Stellar #1*, ed. J. Del Rey. New York: Ballantine Books.

Simmons, G. 2003. *Precalculus mathematics in a nutshell.* Eugene, OR: Wipf and Stock.

Simon, J. 1981. *The Ultimate Resource*. Princeton, NJ: Princeton University Press.

Socher, R., J. Bauer, C. Manning, and A. Ng. 2013. Parsing with compositional vector grammars. In *Proceedings of the 51st Annual Meeting of the Association for Computational Linguistics*, 455–465. Madison, WI: Omnipress.

Socher, R., A. Perelygin, J. Wu, J. Chuang, C. Manning, A. Ng, and C. Potts. 2013. Recursive deep models for semantic compositionality over a sentiment treebank. In *Proceedings of the 2013 Conference on Empirical Methods in Natural Language Processing*, 1631–1642. Stroudsburg, PA: Association for Computational Linguistics.

Stubbs, A. and J. Pustejovksy. 2012. *Natural Language Annotation for Machine Learning*. Sebastopol, CA: O'Reilly.

Sullivan, W., S. Brown, and C. Wetherill. 1978. Eavesdropping: The radio signature from Earth. *Science* 199:377–388.

Sussman, G., and G. Steele. 1998a. The first report on Scheme revisited. *Higher-Order and Symbolic Computation* 11:399–404.

Sussman, G., and G. Steele. 1998b. Scheme: An interpreter for extended lambda calculus. *Higher-Order and Symbolic Computation* 11:405–439.

Tarter, J. 2001. The search for extraterrestrial intelligence (SETI). *Annual Review of Astronomy and Astrophysics* 39:11–48.

Teller, E., Wood, L., and R. Hyde. 1997. Global warming and ice ages: Prospects for physics-based modulation of global change. Paper presented at the 22nd International Seminar on Planetary Emergencies, Erice, Italy, August 1997.

Tesla, N. 1900. Letter to the American Red Cross, New York City.

Tesla, N. 1901. Talking with the planets. *Current Literature* (March): 359–360.

Tesla, N. 1909. How to signal to Mars. *New York Times*, May 23.

Tesla, N. 1919. That prospective communication with another planet. *Current Opinion* (March): 170–171.

Tough, A., ed. 2000. *When SETI Succeeds: The Impact of High-Information Contact*. Bellevue, WA: Foundation for the Future.

Touma, H. 1996. *The Music of the Arabs*. Portland, OR: Amadeus Press.

Townes, C. 1957. Microwave and radio-frequency resonance lines of interest to radio astronomy. *Proceedings of the International Astronomical Union* 4:92–103.

Townes, C. 1983. At what wavelengths should we search for signals from extraterrestrial intelligence? *Proceedings of the National Academy of Sciences USA* 80:1147–1151.

Trauth, K., S. Hora, and R. Guzowski. 1993. *Expert Judgment on Markers to Deter Inadvertent Human Intrusion into the Waste Isolation Pilot Planet*. Albuquerque, NM: Sandia National Laboratories.

USSR Academy of Sciences. 1975. The Soviet CETI program. *Icarus* 26 (3): 377–385.

Vakoch, D. 1998a. Constructing messages to extraterrestrials: An exosemiotic perspective. *Acta Astronautica* 42 (10): 697–704.

Vakoch, D. 1998b. Signs of life beyond Earth: A semiotic perspective. *Leonardo* 31:313–319.

Vakoch, D. 2000a. Cognitive modeling of music perception as a foundation for interstellar message composition. Paper presented at 51st International Astronautical Congress, Rio de Janeiro, Brazil.

Vakoch, D. 2000b. Conventionality of pictorial representation in interstellar messages. *Acta Astronautica* 46:733–736.

Vakoch, D. 2000c. Three-dimensional messages for interstellar communication. In *Bioastronomy 99: A New Era in the Search for Life*, ed. G. Lemarchand and K. Meech, 623–627. Chicago: University of Chicago Press.

Vakoch, D. 2001. Altruism as the key to interstellar communication. *Research News & Opportunities in Science and Theology* 2: 2, 16.

Vakoch, D. 2002. Encoding altruism. *Galileo* 52:30–36.

Vakoch, D. 2004a. The art and science of interstellar message composition. *Leonardo* 37:33–34.

Vakoch, D. 2004b. To the stars, silently. *Leonardo* 37:265.

Vakoch, D. 2006. Describing the probabilistic nature of human behavior in interstellar messages. Paper presented at the 57th International Astronautical Congress, Valencia, Spain.

Vakoch, D. 2007. Bernard M. Oliver's analysis of the role of active SETI strategies in a diversified approach to interstellar communication. Paper presented at the 58th International Astronautical Congress, Hyderabad, India.

Vakoch, D. 2008. Representing culture in interstellar messages. *Acta Astronautica* 63 (5–6): 657–664.

Vakoch, D. 2009a. Anthropological contributions to the search for extraterrestrial intelligence. In *Bioastronomy 2007: Molecules, Microbes, and Extraterrestrial Life*, ed. K. Meech, J. Keane, M. Mumma, J. Siefert, and D. Werthimer, 421–428. Chicago: University of Chicago Press.

Vakoch, D. 2009b. Communicating multiple theories of human psychology in interstellar messages. Paper presented at 60th International Astronautical Congress, Daejeon, South Korea.

Vakoch, D. 2010. An iconic approach to communicating musical concepts in interstellar messages. *Acta Astronautica* 67 (11–12): 1406–1409.

Vakoch, D. 2011a. The art and science of interstellar message composition: A report on international workshops to encourage multidisciplinary discussion. *Acta Astronautica* 68 (3–4): 451–458.

Vakoch, D. 2011b. Asymmetry in active SETI: A case for transmissions from Earth. *Acta Astronautica* 68:476–488.

Vakoch, D. 2011c. A narratological approach to interpreting and designing interstellar messages. *Acta Astronautica* 68 (3–4): 520–534.

Vakoch, D. 2011d. A taxonomic approach to communicating maxims in interstellar messages. *Acta Astronautica* 68 (3–4): 500–511.

Vakoch, D., ed. 2014. *Archaeology, Anthropology, and Interstellar Communication*. Washington, DC: NASA Office of Communications.

Vakoch, D. 2016. In defense of METI. *Nature Physics* 12:890.

Vakoch, D. 2017. Hawking's fear of an alien invasion may explain the Fermi paradox. *Theology and Science* 15 (2): 1–5.

Vakoch, D., T. Lower, B. Niles, K. Rast and C. DeCou. 2013. What should we say to extraterrestrial intelligence? An analysis of responses to "Earth Speaks." *Acta Astronautica* 86:136–148.

Vakoch, D., and M. Matessa. 2011. An algorithmic approach to communicating reciprocal altruism in interstellar messages: Drawing analogies between social and astrophysical phenomena. *Acta Astronautica* 68 (3–4): 459–475.

von Braun, K., and T. Boyajian. 2017. *Extrasolar Planets and Their Host Stars*. New York: Springer.

von Frisch, K. 1967. *The Dance Language and Orientation of Bees*. Trans. Leigh Chadwick. Cambridge, MA: Harvard University Press.

Wang, P., B. Goertzel, and S. Franklin, eds. 2008. *Artificial General Intelligence 2008: Proceedings of the First AGI Conference*. Amsterdam: IOS Press.

Wilkins, J. 1668. *An Essay towards a Real Character and a Philosophical Language*. London: Royal Society London.

Winograd, T. 1972. Procedures as a representation for data in a computer program for understanding natural language. *Cognitive Psychology* 3:1–191.

Wittgenstein, L. 1922. *Tractatus Logico-Philosophicus*. Trans. C. K. Ogden. London: Routledge & Kegan Paul. Originally published as "Logisch-Philosophische Abhandlung," 1921.

Worth, S. 1981. *Studying Visual Communication*. Philadelphia: University of Pennsylvania Press.

Zaitsev, A. N.d. Proposing a METI institute. http://ieti.org/articles/metiinst.pdf.

Zaitsev, A. 2000. One-dimensional radio message for "blind" aliens. Institute of Radio Engineering and Electronics of RAS. http://www .cplire.ru/html/ra&sr/irm/1-D-rm.html.

Zaitsev, A. 2002. Design and implementation of the first theremin concert for aliens. Presentation at the 6th International Space Arts Workshop, Paris, France.

Zaitsev, A. 2006. Transforming SETI to METI. http://www.setileague.org/ editor/metitran.htm.

Zaitsev, A. 2008. Sending and searching for interstellar messages. *Acta Astronautica* 63 (5–6): 614–617.

Zaitsev, A. 2011. METI: Messaging to extraterrestrial intelligence. In *Searching for Extraterrestrial Intelligence: SETI Past, Present and Future*, ed. H. P. Shuch. Berlin: Springer.

Zaitsev, A. 2012. Classification of interstellar radio messages. *Acta Astronautica* 78:16–19.

Zaitsev, A., C. Chafer, and R. Braastad. 2005. Making a case for METI. http://www.setileague.org/editor/meti.htm.

Zaitsev, A., and S. Ignatov. 1999. Broadcast for extraterrestrial intelligence from the Evpatoria deep space center: Report on Cosmic Call 1999. Kotelnikov Institute of Radioengineering and Electronics of RAS. http://www.cplire.ru/html/ra&sr/irm/report-1999.html.

Zhang, Y., B. Lamb, A. Feldman, A. Zhou, T. Lavergne, L. Li, and F. Romesberg. 2017. A semisynthetic organism engineered for the stable expansion of the genetic alphabet. *Proceedings of the National Academy of Sciences* 114:1317–1322.

Zipf, G. 1949. *Human Behavior and the Principle of Least Effort.* Cambridge, MA: Addison-Wesley.

Zipf, G. 1965. *The Psycho-Biology of Language: An Introduction to Dynamic Philology.* Cambridge, MA: MIT Press. Originally published 1936.

INDEX

Almár, Iván, 160
Alpha Centauri, 133
Animal communication
vs. human language, 24, 41–45
information theory, 49–51
as interstellar communication
analog, 20
mathematical ability, 81–82
on the Voyager record, 118
Arecibo message, 14, 100–101,
155
prototype, 1, 140
Arguments against METI,
11–12
shouting in a jungle, 156–161
METI is unscientific, 161–163
METI is wasteful, 163–167
who speaks for Earth, 167–169
Art
as basis of an interstellar
message, 135–137
definition of, 136
Artificial intelligence, 16, 55–56,
59

as basis of an interstellar
message, 17
similarities with extraterrestrial
intelligence, 33
Astraglossa, 12–13, 71–75, 94
design of, 71–74
Astrobot Ella, 16, 55–56
Astrolinguistics, 17, 94, 106–110
definition of, 107
Astronomy, as content of an
interstellar message, 72
Atchley, Dana, 39

Beatles, the, 145
Bernard, Tristan, 6
Bilingual image glossary, 103
Bitmap, 2, 8, 14, 100–101, 103
Boolean logic, 57, 85–86
Breakthrough Initiative, 132–133
British Interplanetary Society,
12
Busch, Michael, 162
Byurakan conference, 19, 56, 58,
99

Calculus of constructions with induction, 109–110
Calvin, Melvin, 39
Cats, 21, 56–57, 141
Cetaceans, 20, 46, 48
Chatbot, 16, 55
Children, as models of extraterrestrial intelligence, 28
Chomsky, Noam, 24, 30, 58, 95
  thesis on animal communication, 41
Cocconi, Giuseppe, 39, 122
Cockell, Charles, 89–90
Cognitive revolution, 29
Cold War, 19
Communication
  fundamental problem of, 51
  relationship to language, 42
Communication Research Institute, 39, 46
Communication with Extraterrestrial Intelligence (CETI), 20
  vs. METI, 20, 22, 25
Computers, as basis of an interstellar message, 21, 57
Concert for aliens, 16
Convergence, 45
Copple, Kevin, 16, 55
Coq, 110
Cosmic Call
  design of, 100–102, 149
  first broadcast, 16, 23, 96, 100
  second broadcast, 16, 55, 103–104
  transmission, 102, 141, 159, 167

Cosmic OS, 64–67
Cros, Charles, 5–6
Cygnus, 16

Darwin, Charles, 42
Davis, Joe, 23, 68–69, 135
DeVito, Carl, 15, 75, 78–79
Digital encoding, 124–126
Discrete infinity, 28
DNA
  as executable code, 68–69
  in interstellar messages, 14, 68–70, 101, 140
  as universal feature of life, 67–70, 90–91
Dolphinese, 46
Dolphins
  communication system, 41, 43, 47, 50–51
  experiments on, 39
  as model of extraterrestrial intelligence, 40–41, 43
Doyle, Laurance, 51–53
Drake, Frank
  Arecibo message, 14, 140, 155
  Green Bank conference, 37–38
  Pioneer plaque, 112–115
  Project Ozma, 13, 20, 37
  prototype interstellar message, 1–3, 162
  Voyager records, 118
  Waste Isolation Pilot Plant task force, 138–140
Drake-Helou limit, 121
Druyan, Ann, 15

Dumas, Stephane, 16, 100–103
Dutil, Yvan, 100–103, 149

Economy, as universal principle,
    33–35
EISCAT, 145
Elliot, James, 20–21, 25
Elliot, John, 60–61, 63
Embodied mathematics
    definition of, 81
    grounding metaphors, 82–83
    relationship to set theory, 85–87
Encephalization, 45
Epsilon Eridani, 37–38, 135
Evolution, 45–46, 89–91
Evpatoria radio telescope, 16,
    100–101, 103, 127, 157
Exoplanet, 22, 74

Fitzpatrick, Paul, 64–67
Flammarion, Nicolas, 6, 128–129
Freudenthal, Hans, 13, 56, 94–99,
    104–106, 108

Galilei, Galileo, 23, 77
Galton, Francis, 7–8
Gauss, Carl, 4, 139
Generative grammar, 29, 58, 95
Gödel, Kurt, 80
Godwin, Frances, 4
Green Bank conference, 1,
    38–40, 43
    attendees, 39

Heliotrope, 5
Herzing, Denise, 46–48

High Resolution Microwave
    Survey, 164
Hockett, Charles, 24
Hogben, Lancelot, 12–13, 71–74,
    94
Hoyle, Fred, 57
Huang, Su-Shu, 39
Human language chorus corpus,
    60
Hydrogen line, 37

Icons
    as basis of an interstellar
        message, 140
    for representing music, 152
Images, conventionality of,
    139–140
Indeterminacy of reference, 26–28
Intelligence filter, 51, 53
Interstellar message
    first broadcast, 14
    first musical broadcast, 16
    first prototype, 1–2, 140
    first scientific broadcast, 16
    how to distinguish from natural
        phenomena, 52, 120
    symbolic messages, 13
Intuitionism, 81

Jansky, Karl, 12

Lambda calculus, 17, 64, 80, 109
Language corpora
    annotation, 62–63
    content, 60
    ideal size, 61

Lasers
  encoding information, 130–131
  energy requirements, 131
  interstellar communication, 130
  invention, 129
Leibniz, Gottfried, 84–85
Lilly, John, 39, 41–43
Lincos, 13–14, 16, 17, 56, 57, 80,
    94, 111, 142
  criticism, 104–106, 108
  description of the system, 96–97
  inspiration for the Cosmic Calls,
    100–101
  second generation, 106–110
Loebner Prize, 55
Logic
  as basis of an interstellar
    message, 17
  constructive, 109
  predicate, 108–109
  propositional, 108
LSD, 40, 46

Magic frequencies, 37, 122–123
Marconi, Guglielmo, 8, 10–11
  alleged interplanetary
    transmissions, 10
Markov process, 58
Mars, 6–11
  depiction in interstellar
    messages, 177
Martians, 6, 8, 31, 73, 128–129
Mathematical formalism, 79–81,
    98
Mathematical Platonism, 78, 80,
    86, 88

Mathematics
  as basis of an interstellar
    message, 11, 13, 33, 72, 77, 97
  theories of, 77–79
McCarthy, John, 33
Messaging Extraterrestrial
    Intelligence (METI)
  origin of the term, 23
  premodern, 3–11
  symbolic turn, 12–19
Metalanguage, 17, 109
METI International, 145
Microwave band, 119
Microwave window, 119–120
Millstone radar, 23, 135
Minimalist program, 30
Minsky, Marvin, 21, 33–35,
    56–58, 92
Mirrors, as communication
    medium, 5–6, 8–9, 129
Moro, Andrea, 31
Morrison, Philip, 39
Morse code, 6, 8
Moving pictures, as basis of an
    interstellar message, 142–144
Music
  as basis of an interstellar
    message, 16, 145–153, 168
  interstellar transmission, 145
  relationship to human
    cognition, 150–152

NASA, 112, 122, 132, 145, 164
Natural language
  vs. animal communications,
    41, 44

as basis of an interstellar
  message, 25–26
faculty of, 30–31
Hockett's design features, 24
as means of expressing thought,
  24, 41
Natural language processing,
  58–60, 109
Neurolinguistics, 30–32
Nonhuman primates, 43, 45–46,
  48
Nuclear weapons, as interstellar
  beacon, 20

Oehrle, Richard, 15, 75, 78
Oliver, Barney, 2, 39, 162
Ollongren, Alexander, 17, 63,
  80–81, 94, 106–110, 142, 153
Optical METI, 128–133
  early proposals, 4–5
Optical SETI, 132
Order of the Dolphin, 40–41,
  130

Pearman, J. P., 39–40
Pioneer plaque, 15, 101, 112–116,
  167
Prix Guzman, 7
Project Cyclops, 122
Project Ozma, 1, 13, 20, 37–38
Pulsar, 42
  as clocks, 114
  information content, 52–53
Purcell, Edward, 38

Quine, W. V. O., 26–28

Radio
  astronomy, 12
  early experiments, 8–11
  modulation, 123–124
Radioglyph, 73
Reddick, Rachel, 162–163
RuBisCO, 68–69
Russell, Bertrand, 98–99

Sagan, Carl
  Byurakan conference, 19–20,
    99
  design of Pioneer plaques,
    14–15, 112–116, 167
  design of Voyager record,
    14–15, 167
  on dolphins, 46
  Green Bank conference,
    39
  Waste Isolation Pilot Plant task
    force, 138–139
Sagan, Linda, 112
San Marino index, 160–161
Scheme, 64
Schiaparelli, Giovanni, 7
Science
  as basis of an interstellar
    message, 16, 75–76
  METI's status, 161–163
Search for Extraterrestrial
  Intelligence (SETI)
  comparison with METI,
    22
  origins, 38
  in the USSR, 19, 56, 99
Self-interpretation, 13, 110

Set theory, 85–87
Shannon entropy, 50–51, 53
Shklovskii, Iosif, 19
Silverberg, Robert, 25
Sónar message, 148–150
Starfish Prime, 21
Statistical revolution, 59
Struve, Otto, 39

Tabula rasa, language acquisition
    hypothesis, 28–29
Tau Ceti, 37–38, 135
Team Encounter, 100–101, 103
Teen Age message, 104, 124
    design of, 146–147
Tesla, Nikola, 9–10
    belief in extraterrestrial
    communication, 9
Theremin, 147
Townes, Charles, 129

Universal grammar
    experiments with brain imaging,
        31
    theory of, 29
    violation of, 31

von Littrow, Joseph, 5, 128
Voyager golden record, 15, 135,
        138, 141–142, 145
    contents, 116, 118–119
    criticism, 118

Waste Isolation Pilot Plant,
        137–139
"Water hole," 122

Zaitsev, Alexander, 23, 102,
    145–146
Zipf's law, 49–52, 63